IEE Telecommunications Series 13

Series Editors: Prof. J.E. Flood
C.J. Hughes

Advanced signal processing

Previous volumes in this series

Volume 1	Telecommunications networks J.E. Flood (Editor)
Volume 2	Principles of telecommunication-traffic engineering D. Bear
Volume 3	Programming electronic switching systems M.T. Hills and S. Kano
Volume 4	Digital transmission systems P. Bylanski and D.G.W. Ingram
Volume 5	Angle modulation: the theory of system assessment J.H. Roberts
Volume 6	Signalling in telecommunications networks S. Welch
Volume 7	Elements of telecommunications economics S.C. Littlechild
Volume 8	Software design for electronic switching systems S. Takemura, H. Kawashima, N. Nakajima Edited by M.T. Hills
Volume 9	Phase noise in signal sources W.P. Robins
Volume 10	Local telecommunications J.M. Griffiths (Editor)
Volume 11	Principles and practice of multi-frequency telegraphy J.D. Ralphs
Volume 12	Spread spectrum in communication R. Skaug and J.F. Hjelmstad

Advanced signal processing

Edited by D.J.Creasey

Peter Peregrinus Ltd on behalf of The Institution of Electrical Engineers

Published by: Peter Peregrinus Ltd., London, UK.

© 1985: Peter Peregrinus Ltd.

All rights reserved. No part of this publication may be reproduced, stored in a retrieval system or transmitted in any form or by any means — electronic, mechanical, photocopying, recording or otherwise — without the prior written permission of the publisher.

While the author and the publishers believe that the information and guidance given in this work is correct, all parties must rely upon their own skill and judgment when making use of it. Neither the author nor the publishers assume any liability to anyone for any loss or damage caused by any error or omission in the work, whether such error or omission is the result of negligence or any other cause. Any and all such liability is disclaimed.

British Library Cataloguing in Publication Data

Advanced signal processing
 — (IEE telecommunications series; 13)
1. Signal processing
I. Creasey, D.J.   II. Series
621.38'043   TK 5102.5

ISBN 0-86341-037-5

Printed in England by Short Run Press Ltd., Exeter

Contents

	Page
Editorial	xi
1 Transmitter aerials	1
1.1 Introduction	1
1.2 The function of radiating devices	1
1.3 Electrical to physical conversion	2
1.4 Direct systems	3
1.5 Maximising information gained	3
1.6 Observation time	4
1.7 Transfer of energy to the propagating medium	4
1.8 Conclusions	6
2 High power amplifier design for active sonar	8
2.1 Introduction	8
2.2 Applications	8
2.3 Constant-power amplifier implementation for surveillance sonars	10
3 Radar transmitters	15
3.1 Introduction	15
3.2 The transmitter as a power converter	15
3.3 The transmitter as a signal synthesiser or signal amplifier	19
3.4 The effects of defects of transmitter signal quality	21
3.5 The measurement of transmitter signal quality	22
4 Receiver array technology for sonar	23
4.1 The problem	23
4.2 The sonar array	23
4.3 The hydrophone	24
4.4 The baffle	24
4.5 Flow noise and domes	24
5 New underwater acoustic sensors	26
5.1 Introduction	26

	5.2	The optical fibre hydrophone	26
	5.3	Composite materials	30
	5.4	Multilayer transducers	32
6	Diversity techniques in communications receivers		35
	6.1	Introduction	35
	6.2	Envelope properties	36
	6.3	Bit error rate performance in fading conditions	37
	6.4	Diversity reception	40
	6.5	Selection diversity	43
	6.6	Maximal ratio combining	43
	6.7	Equal gain combining	45
	6.8	Improvements obtainable from diversity	45
	6.9	The effect of diversity on data systems	46
7	GaAs IC amplifiers for radar and communication receivers		50
	7.1	Introduction	50
	7.2	Low frequency amplifiers	51
	7.3	Low noise X-band radar amplifier	54
	7.4	Broadband amplifiers for EW and signal processing	56
	7.5	A comparison of circuit techniques	59
	7.6	Conclusions	60
8	Integrated optical techniques for acousto-optic receivers		62
	8.1	Introduction	62
	8.2	The optical waveguide	63
	8.3	The SAW transducer	64
	8.4	Input and output coupling of the laser beam	65
	8.5	Lenses	66
	8.6	Dynamic range and frequency resolution	67
	8.7	Discussion and conclusions	68
9	Logarithmic receivers		72
	9.1	Introduction	72
	9.2	Types of logarithmic amplifier	72
	9.3	Achieving a logarithmic response	73
	9.4	Circuit considerations	77
	9.5	Integrated circuits	78
	9.6	Future integrated circuits	80
10	CCD processors for sonar		81
	10.1	Introduction	81
	10.2	Time domain programmable spatial and signal processing	82

	10.3	Implementation of programmable spatial and signal processing	84
	10.4	Programmable beam-forming and steering devices	85
	10.5	Programmable signal processing devices	86
	10.6	Conclusions	88
11	Learning from biological echo ranging systems?	91
12	Acousto-optic correlators	93
	12.1	Introduction	93
	12.2	Optical components	93
	12.3	One-dimensional a.o. correlators	95
	12.4	Two-dimensional correlators	97
	12.5	Conclusions	102
13	GaAs MESFET comparators for high speed Analog-to-Digital conversion	103
	13.1	Introduction	103
	13.2	Principle of GaAs FET comparator	104
	13.3	IC fabrication	108
	13.4	Experimental results	108
	13.5	Conclusions	109
14	Designing in silicon	110
	14.1	Introduction	110
	14.2	Semi-custom IC design	111
	14.3	Custom LSI design	114
	14.4	Computer aided design tools	116
	14.5	The user's point of view	116
15	Very high performance integrated circuits	118
	15.1	Introduction	118
	15.2	VHPIC process technology	120
	15.3	Defence applications of VHPIC	121
	15.4	VHPIC as a cost reducing technology	125
	15.5	Civil applications of VHPIC	126
	15.6	VHPIC design	127
	15.7	Conclusion	130
16	Very high speed integrated circuits (VHSIC) technology for digital signal processing applications	132
	16.1	Introduction	132
	16.2	Overview of the VHSIC program	132
	16.3	Signal processing applications of VHSIC	138

	16.4	Design automation	141
	16.5	Future digital signal processing advances	141
17	Digital filters	143
	17.1	Introduction	143
	17.2	Structures	144
	17.3	Design of infinite impulse response filters to meet piecewise-constraint magnitude-vs-frequency specifications	146
	17.4	Design for general magnitude-vs-frequency specifications	147
	17.5	Implementation	149
18	Display types	153
	18.1	Introduction	153
	18.2	Cathode ray tube	154
	18.3	Vacuum fluorescent flat-panel displays	155
	18.4	Electroluminescent flat-panel displays	156
	18.5	Gas discharge (plasma) flat-panel display	157
	18.6	Liquid crystal display	159
	18.7	The future	160
19	Scan converters in sonar	162
20	Display ergonomics	170
	20.1	Introduction	170
	20.2	Existing evidence	171
	20.3	Ambient lighting	172
	20.4	Perceptual structuring	174
	20.5	Visual coding	175
	20.6	Two specific problems	175
	20.7	The provision and use of information	176
	20.8	Failures and errors	177
	20.9	Measurement	177
	20.10	Résumé	178
21	Simulators	180
	21.1	Introduction	180
	21.2	Simulators: aids to advanced signal processing	180
	21.3	Dynamic electromagnetic environment simulators	183
	21.4	Field portable trials simulators	187
22	High throughput sonar processors	191
	22.1	Introduction	191
	22.2	Technology background	192
	22.3	System requirements	193

	22.4	System architectures	195
	22.5	Macro development	197
	22.6	Algorithm development	199
23	Optical fibre systems for signal processing		202
	23.1	Introduction	202
	23.2	Components for fibre optic signal processing	204
	23.3	Some examples of fibre optic signal processing	205
	23.4	Some basic limitations on all optical monomode fibre processors	206
	23.5	Future trends and possibilities	207
24	Satellite communications		209
	24.1	Background	209
	24.2	Satellite system operation	209
	24.3	System studies and link budgets	210
	24.4	OLYMPUS	211
	24.5	Advanced modulation techniques	214
	24.6	The future	216
25	A VLSI array processor for image and signal processing		218
	25.1	Introduction	218
	25.2	System architecture	218
	25.3	Programming	219
	25.4	Applications	219
26	VHPIC for radar		221
	26.1	Introduction	221
	26.2	Radar processing functions	221
	26.3	Processes and performance requirements	225
	26.4	VHPIC technology	227
	26.5	Architecture and technology integration issues	229
	26.6	Conclusions	230
27	High performance and high speed integrated devices in future sonar systems		232
	27.1	Introduction	232
	27.2	Function unit architecture	233
	27.3	Sonar system architecture	233
	27.4	Building blocks to realise function units	234
	27.5	VLSI implementation	238
28	Ada		242
	28.1	Introduction	242

Contents

28.2	Aspects of the language	242
28.3	Ada availability	246

29 Future for cryogenic devices for signal processing applications 248
    29.1 Introduction 248
    29.2 Analog signal processing 249
    29.3 Digital signal processing 253
    29.4 Hybrid signal processing 255
    29.5 Discussion 255

30 Application of the ICL distributed array processor to radar signal processing 257
    30.1 Introduction 257
    30.2 Historical background 257
    30.3 Requirements for signal processing for agile platform airborne radars 260
    30.4 The military DAP solution 261
    30.5 Future development 263

31 Advanced image understanding 265
    31.1 Introduction 265
    31.2 Declarative methods in unconstrained image understanding 266
    31.3 The RSRE 'ASARS' image understanding system 269

32 Advanced signal processing aspects of automatic speech recognition 275
    32.1 Introduction 275
    32.2 Stages in speech communication − from mind to mind 275
    32.3 Levels of description − from waveform to meaning 276
    32.4 Speech signal processing − from waveform to pattern 277
    32.5 Speech pattern processing − from patterns to words 277
    32.6 Future developments − from 1984 to 1994 279

33 VLSI architectures for real-time image processing 282
    33.1 Introduction 282
    33.2 Image processing dataflow and computational structures 283
    33.3 Criteria for VLSI implementation 285
    33.4 Candidate VLSI architectures for real-time image processing 286

Index 294

Editorial

D. J. Creasey
Department of Electronic and Electrical Engineering, The University of Birmingham, P.O. Box 363, Birmingham, UK

The International Specialist Seminar on "ADVANCED SIGNAL PROCESSING" is the fourth in a series of specialist seminars. The earlier seminars have been spaced at intervals of three years and they have all been held in Scotland. Thus the 1984 Seminar held at the University of Warwick in England broke the Scottish tradition of these events. The three previous seminars held in 1973, 1976 and 1979 had the titles "The Component Performance and Systems Applications of Surface Acoustic Wave Devices", "The Impact of New Technologies in Signal Processing" and "Case Studies in Advanced Signal Processing" respectively. In September 1983 the IEE organised a similar International Specialist Seminar on "Systems on Silicon". This took place at Stratford on Avon and resulted in the 1984 publication by Peter Perigrinus of the book "Systems on Silicon" which was edited by P. B. Denyer.

Since the last signal processing seminar there have been many advances in signal-processing technology. A number of the professional groups within the Institution of Electrical Engineers regarded the five-year gap as being too long and so preparations were begun in June 1983 for this fourth seminar. An organising committee was set up by the IEE under the chairmanship of Reg Humphries and the Seminar was organised in association with the IEEE and the IERE. The Specialist Seminar covered the following topics:

Transmitters, receivers and sensors,
Information processors,
Display processors,
Future systems, and
Case studies of future systems.

The intention of the Seminar was to bring together component specialists and systems designers to continue their dialogue at an international level in the spirit of the previous seminars.

The Organising Committee intended that a book should be produced as a record of the papers presented at the Seminar.

The Committee also intended that the book should be available to delegates before the start of the Seminar. Inevitably, as with all well laid plans, things did not go exactly as was hoped and the publication had to be postponed until after the Seminar.

It is obvious that signal-processing schemes need adequate components for implementation. As algorithms increase in complexity, the signal-processing engineer continually looks for faster and more reliable components. At the front end of any system the engineer must interface his system to some form of transduction unit. These units can often place limits on the performance of the signal processor. At the other end of the system, the information gleaned by the transduction elements and processed into an acceptable form must be presented to an observer through some form of display. This book contains chapters covering the many aspects of modern signal-processing systems from the transduction unit through to the display. Each chapter is written by an expert in the appropriate area. Some chapters are deliberately intended to be of a tutorial nature while others contain material from research laboratories which has not been published previously.

The major subject areas covered include radar, sonar and communications systems. As I have read the various chapters, I have been struck by the similarity of the problems faced by designers in these three areas. Often, because of the different frequencies involved and the different environments in which the systems have to operate, detailed solutions may appear to be different. Nevertheless signal-processing engineers from different disciplines can always learn from one another.

Recent developments in signal-processing components have given the designer a new dimension in which to work. Components have become smaller and faster and with increased versatility. We are constantly reminded that the reduction in size with consequent increase in speed cannot continue indefinitely. We are told by the component manufacturers that they are reaching fundamental limits set by the wavelength of light and the structure of the crystal lattice. Yet these same manufacturers continue to improve their products. Is it any wonder that system designers ask for more of the same thing? Of great importance recently have been the VHSIC and VHPIC programmes set up in the US and UK respectively. Designers await with eagerness the time when components from these programmes are available for all to use.

We must not be blinded by the glamour of the success of fast modern digital microelectronics. Both interface components and alternative technologies are striving to keep up with the digital pace setters. You will find within the pages of this book sufficient evidence that the digital processor is still not king in every sphere. Indeed, as clever as we are, man-made signal-processing engines still fall far short of those prossessed by many animals. We have much to learn from the operation of our own brains in such things as pattern recognition, classification, learning and intelligence. Computing machines can outperform us in simple tasks such as number-crunching but they have a long way to go before they become generally superior to many biological systems.

Much of this book deliberately concentrates on the component aspects of modern signal processing. However the Organising Committee believe that in the Seminar programme, the balance between components, systems, architecture etc. was correct. It was intended that the Seminar should reflect changes in signal processing that are being brought about by components which were only pipe dreams a few years ago. As this book is a record of that Seminar it also reflects the importance of components.

David Creasey, The University of Birmingham, August 1985.

Chapter 1
Transmitter aerials

M. J. Withers

ERA Technology Ltd., Leatherhead, UK

1.1 Introduction

So often there is great difficulty in choosing a title which concisely conveys the subject being considered. In this instance, the problem is compounded since particular words convey different meanings amongst specialist groups and, similarly, common processes are often named differently from one discipline to another.

The subject under consideration has all the above problems and, in addition, is extremely diverse. Radar has its own terms and unique antenna solutions to match particular requirements, while sonar has a quite different set. Radio frequency communications is sub-divided with 'aerials' and 'antennas' being used depending upon which section of the electromagnetic spectrum is being utilised. Similar sub-divisions occur with sonar, seismology, radio astronomy, etc.

It is not until one attempts to write a short exposition for specialists having a wide spectrum of interests that it becomes fully apparent how diverse and ingenious man has become in exploiting nature to his benefit.

To keep the task within bounds I have concentrated on the basic function to be performed by a radiating device, whether it be aerial, antenna, array . . . or transducer, and to consider its effect upon a system whose function is to gain information.

Most of this chapter concentrates on transmitter antenna applications in radar and radio communications. However, where there is relevant equivalence in other fields, such as sonar, it is discussed so as to draw the reader's attention to the similarities, or lack of them.

1.2 The function of radiating devices

Radiating devices are basically required to provide an efficient means of converting an electrical effect into a physical effect. In many instances this is only

2 Transmitter aerials

part of its role as it may also be required to convert from physical to an electrical effect; that is to operate in a receiving mode. This dual operation often leads to a compromise design, unless space, cost, etc. allow separate devices to be employed.

The characteristics generally expected, depending upon the particular system application, are as follows:

(a) Efficient conversion of electrical signal into propagating wave in medium or vice versa.
(b) To direct efficiently, or collect, the propagating wave to maximise the overall signal-to-noise ratio (or signal-to-interference ratio).
(c) To maximise the information gained, in the true sense, and therefore to minimise the uncertainties/ambiguities.
(d) To collect information in a period of time compatible with the duration of the events being observed.
(e) To ensure adequate energy is transferred to the propagating medium without breakdown to allow the above requirements to be achieved.
(f) The application of the most cost-effective solution.

Each of the above requirements will be discussed in detail and the interaction between the sensor characteristic and the systems design will be explored. Within this section the emphasis will be on the transmission demands but the choice of receive characteristic should always be borne in mind since they are an important degree of freedom in making the final selection of the system to be adopted.

1.3 Electrical to physical conversion

The majority of systems struggle to achieve adequate signal-to-noise ratio. The system equations can be arranged to demonstrate that it is the energy radiated in a given direction that sets the limit of performance. Transmitter power is not a commodity in plentiful supply and one cannot afford to squander it through poor design of the radiating device. Furthermore, lack of conversion efficiency on receive is wasteful of spectrum and leads to interference for other users. At worst it can enhance the chance of intercept by an undesired third party.

Therefore, conversion efficiency is fundamental but the phenomena used in the radiating devices are often not efficient in themselves, for example in sonar, the piezoelectric effect. To make the conversion moderately efficient it is necessary to make the device resonate mechanically and to devise an electrical matching network which is also usually resonant. The total ohmic losses compared with the radiation resistance limit the efficiency achievable. Very high power applications give rise to the problem of heat dissipation in the transducer.

The need to 'tune' the sensor places a further limit on the system sensitivity that can be achieved since it sets an upper bound on the amount of pulse-compression that can be employed.

1.4 Direct systems

Sensitivity in some systems is markedly increased by introducing directivity to concentrate the energy, Reintjes and Coate (1), when applied equally on transmission and reception it provides the best system improvement but if the mode of operation involves an angular search to locate targets then there is no net gain if detection has to be achieved within a given time period, Skolnik (2). The reason is that, for a given sector, a given amount of energy is required to make a detection at a specific range. Narrowing the transmission beam by a factor 'n' means that the beam has to scan the sector and although the incident power density on the target is greater by 'n' it can only dwell for '1'/'n' of the time if the sector is to be explored over the same total time duration. A similar argument applies to the receiving mode. Although the detection capability has not been enhanced, it is possible to determine more precisely the angular direction of the target and also to assess whether there are up to 'n' targets present at a given range.

This argument is not true for all situations due to a variety of reasons, mainly depending upon the sources of noise in the system and the non-linear characteristics of the detection process in the low signal-to-noise ratio condition, Schwartz, Bennett and Stein (3).

Directivity is improved by increasing the linear dimension of the radiator; this can be achieved in a variety of ways, such as simply using a larger aperture transducer or coupling together an array of transducers. Another technique is to use a shaped reflector to focus energy onto the transducer. Hybrid systems utilising a combination of reflector and an array in the focal region are also possible.

1.5 Maximising information gained

The directional properties and resolution capabilities depend upon how the energy is distributed across the radiating aperture. It is well known that the maximum gain will be achieved when a uniform aperture illumination is employed. However, the sidelobe levels which occur are high and this gives rise to serious ambiguity problems. To minimise the effect it is usual to taper the amplitude distribution across the aperture, but this leads to lower gain and poorer resolution.

An alternative solution is to reconfigure the aperture illumination in a way such that it is still uniform but that the position of the sidelobes, and the associated nulls, are moved in directional position relative to the main beam. One of the simplest ways of doing this is to switch out a few elements at the end of the array. This moves the sidelobe nulls and peaks in angle. By noting the response of the system for a number of specified configuration changes it is possible to resolve the ambiguities and deduce the target distribution.

Sources of directional interference may be present and then the ability to place nulls to maximise the signal-to-interference ratio becomes a requirement. There are many techniques available and are described by Davies (4).

1.6 Observation time

Some events last for a period of time which is long compared with the observation time, others only provide a 'fleeting glimpse' and the maximum amount of information has to be extracted in a given time. Often the quality of the information will be poor because of the limited time available and the number of mathematical variables to be determined simultaneously.

In many instances, not all the variables have to be determined over one short period. For example, in a search radar it is important to determine rapidly the arrival of all new targets within the coverage area since they may be a potential threat. Once having made the detection, the next requirement is to determine the level of threat. If this is found to be low then these targets do not need to be considered very often. This then allows more time to explore in greater detail those targets of high threat in order to place them in ranking order. There will be some objects present that provide totally unwanted responses, usually referred to as clutter, reverberation, etc. Spending time exploring these is a total waste, and ways of minimising, or eliminating these by the configuration of the sensors should always be attempted.

1.7 Transfer of energy to the propagating medium

To perform the various tasks outlined, it is obvious that a certain amount of energy, and hence power, is required to be converted into a propagating wave. Due to the constraints placed on the system, it is often necessary to use high energy levels. The first problem is to find an emission source with adequate power generating capability. The next problem lies within the antenna/transducer to ensure that it will handle the power levels involved without operating in a non-linear manner, overheating or breaking down.

The final problem is the propagating medium itself, whether it be breakdown by arc creation, or cavitation in sonar, as the primary difficulty or non-linearities, dispersion, etc., they all seriously limit system capabilities.

To overcome the problems, methods of distributing the power have been devised, such as time spreading (pulse expansion/compression) or spatially spreading (arrays).

1.7.1 *Pulse expansion/compression*
General limits occur on the amount of expansion/compression possible due to the following factors:
− time stability of the propagating medium

- relative velocity of the target and/or medium
- attentuation with frequency of medium (propagation windows)
- anomalous propagation
- permitted (allocated) bandwidth of transmission
- ambiguities

Usually the most severe limit is set by the bandwidth of the radiator. The element itself has a finite bandwidth, as mentioned in an earlier section, but in addition any array has an inherent bandwidth due to its transient response related to its physical dimensions.

Consider a linear array of elements with a short burst of carrier applied to each element at the same time; the time duration of the burst being made numerically equal to the time for a signal to propagate from one element to the next along the array. At a distance from the array, in the broadside direction, each burst will be time and phase coincident and so a single short pulse will be observed. In the end-fire direction the pulse will appear radiating in sequence which will merge into a long single pulse, related to the propagation time along the array. It has therefore been demonstrated that the array has different transient behaviour, and hence bandwidth capabilities depending upon the spatial direction of radiation.

1.7.2 Spatially spread (distributed) transmission

To reduce the field intensity in front of the transmission elements to avoid breakdown requires the power to be distributed to a number of elements. One such method is shown in Figure 1.1. This may be done from a single transmitting

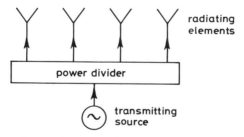

Fig. 1.1. *Distribution of power from a single source to radiating elements*

source by means of a distribution network. Alternatively, each element in the array can have its own source. In the latter case the sources may either be controlled to radiate all in phase (as shown in Figure 1.2a) or to radiate asynchronously but having the phase 'remembered' or reconstructed (Figure 1.2b). The first example provides a single beam in space, which can be phase steered. The second example is effectively a multiple beam transmission with the average power distribution across the sector remaining uniform since each source has random phase.

If the instantaneous phases of the sources are 'remembered' then it is possible to reconstruct the instantaneous phased sum for the receiver processor. That is,

6 Transmitter aerials

the instantaneous direction of maximum response is calculated. The advantage this has is that with the right beam-forming networks (those using time delays) it is possible to overcome the transient bandwidth problem (4).

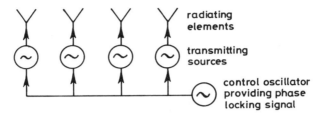

Fig. 1.2a. *Distributed power sources — coherent radiation*

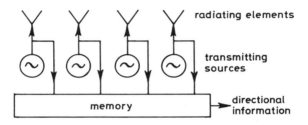

Fig. 1.2b. *Distributed power sources — asynchronous radiation*

There are a number of configurations for these directionally decorrelated transmitter beam systems, but some would require high power beam-forming networks while others lead to concentrated field intensities close to the antenna system. These could lead to breakdown of the propagating medium.

The basic block diagram of a system which overcomes the above problems is shown in Figure 1.3. An essential part of the process is that the transmit and receive arrays operate through beam-forming networks and correlation receivers.

Transmitted beam decorrelation can be obtained by:

(a) time separation of element excitations
(b) frequency separation of element excitations
(c) code separation of element excitations

The merits of the different systems have to be assessed against particular applications since they may give rise to an inordinate amount of storage requirement, or an unacceptable bandwidth occupancy.

1.8 Conclusions

An attempt has been made to outline the multitude of interrelated constraints that arise in devising radiating systems.

Considerable effort is being expended in all fields to improve the performance of traditional radiating devices, whether they be radar, satellite or communication antennas on the one hand or sonar transducers on the other hand.

The development of higher power sources, together with more severe system demands, mean that breakdown problems of radiator, or medium, have to be carefully considered and ways found of overcoming the difficulty.

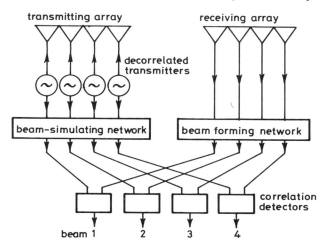

Fig. 1.3. *Schematic transmit/receive system for directionally-decorrelated transmitted beams*

One solution is to use array signal processing and now that much greater capacity in terms of storage capacity exists for numerical processing, it is reasonable to explore new system configurations previously considered too complicated to construct.

Furthermore, these systems can be made adaptive to improve the information efficiency for a given radiated power. To fulfil the greatest system potential it is necessary to match the possible adaptive capability on transmission with even greater sophistication on reception.

1.9 References

1. REINTJES, J. F., and COATE, G. T., 1952, 'Principles of Radar', McGraw-Hill.
2. SKOLNIK, M. I., 1969, 'Radar Handbook', Chapter 2.
3. SCHWARTZ, M., BENNETT, W. R., STEIN, S., 1966, 'Communication Systems and Techniques ', McGraw-Hill, Chapter 3.
4. RUDGE, A. W. R., MILNE, K., OLVER, A. D., and KNIGHT, P., 1983, 'The Handbook of Antenna Design', Peter Peregrinus, 2, Chapter 13, D. E. N. Davies.

Chapter 2
High power amplifier design for active sonar

P. D. Franks
Plessy Marine Ltd U.K.

2.1 Introduction

The definition of "Active Sonar" is applied to systems with a wide range of operating parameters. Sonar power amplifiers operate in environments ranging from large shipborne equipments housed in electronics cabinets to small, self-contained expendable sonobuoys. Duty cycles of operation can range from high-resolution sonars, which deliver short, high frequency pulses with low duty cycle up to systems which are required to operate continuously, often over a very wide bandwidth. Additionally, control sophistication ranges from simple single-frequency generators to modules that deliver complex signals to closely defined tolerances in amplitude over a range of frequency.

This paper presents an overview of the application areas and their specific requirements. A more detailed examination of active surveillance-sonar parameters leads to a number of options for power amplifier design. Finally, the principles developed are demonstrated in design examples for power amplifiers.

2.2 Applications

Generally, the different applications will lead to a range of different hardware configurations. The common requirements are small size and high conversion efficiency. Usually, "high fidelity" reproduction, as in audio amplifiers, is not a requirement.

2.2.1 *High-resolution amplifiers*

It is in this application that conversion efficiency has a reduced importance. Typically, transmission pulse lengths are in the range 10^{-4} seconds, with duty cycles of $< 1\%$.

Pulse amplitude and phase are of importance in multielement array systems where a transmit beam can be accurately steered.

The energy for the transmitted pulse is stored in a capacitor which is "trickle charged" from the mains supply, thus avoiding peak-power demands during the transmit period. Typically the capacitor will discharge 20% of its energy on transmission. The lack of thermal constraints can lead to a very compact implementation. A push-pull class-B output stage transformer coupled to a transducer, is employed, thereby avoiding additional reactive components normally associated with class-D operation. A shaped envelope of phase split sine wave is applied to the output stage. At the start of transmission the impedance presented by the load will be low, so that soft start or current limiting is employed to limit the rate of energy supplied to the load. In what is, basically, a linear amplifier, accuracy of the amplitude of pulse may be optimised by applying negative feedback to each stage.

2.2.2 Low frequency surveillance amplifiers
This type of power amplifier has more in common with switched-mode power supply design than audio-amplifier design. Duty cycles are typically up to 20%, so that long-term thermal dissipation will have an impact on the size and layout of the module. Also, the modern trend in sonar is for longer pulses with more closely defined characteristics in frequency and amplitude. The thermal time constant of semiconductors is such that they have to be rated for average power during the pulse, with little or no derating for the duty cycle. Longer time-constant components, such as inductors, can however be designed for the lower long-term thermal average with considerable saving in size and weight.

A typical design will consist of two stages. The first stage performs mains rectification and voltage regulation to provide a stable d.c. supply. This is followed by an output inverter which is usually in a bridge configuration (Reference 1), driven in class-D at the output frequency. The bridge may be configured to be voltage or current driven.

The mode of control exerted on the output is an important consideration. Constant-current, constant-voltage or constant-power configurations are possible. In the majority of cases, constant-power operation is to be preferred. The advantages are:

(a) The sonar transducer load presents a wide range of admittances over a typical transmission bandwidth. In constant-current or constant-voltage modes, the transmitted amplitude could vary significantly with frequency.

(b) In arrays consisting of elements less than a wavelength apart, the load admittance is altered by mutual coupling between adjacent elements. In some arrays with individual elements driven in constant voltage, the interactions can be large enough to cause some elements to sink rather than source power! The effect on beam patterns in general can be severe. If each element is driven with constant power, these interaction effects are minimised.

(c) Constant-power operation leads to well defined operating levels and stresses in the components in the power amplifier.

(d) With constant power, the load on the mains supply is more consistent, leading to a full, optimum use of mains power and minimal size for components such as mains filters.

Constant-power operation has the disadvantage of requiring more complex control circuitry, though this is a small price to pay for the benefits obtained. The structures of constant-power amplifiers are expanded in the next section.

2.2.3 *Continuous wideband amplifiers*

These will often have similar architecture to the low-frequency surveillance amplifiers. In applications such as expendable decoys, conversion efficiency, size and weight are of critical importance. Such amplifiers will be battery driven, so that their power supply has inherently good voltage regulation. Once again, for the advantages described previously it is desirable to try to achieve constant-power operation, and this may be implemented by pulse-width modulating the drive waveforms to the inverter stage such that the current drawn from the battery is constant. Typically the load connected to the inverter will consist of several transducers, covering a range of frequency bands, connected via some cross-over arrangement similar to that in an audio loudspeaker system.

To achieve constant power, the control system has to track changes in load impedance caused by the rapid rates of change of frequency in a wideband drive situation. If a large bandwidth is required over a short integration time, constraints in the feedback control stability will not allow accurate control to be maintained for the rapid accelerations of phase in the drive signal. In this case the control system will be designed to seek a defined power level in the long term and the system design has to adjust the statistics of the drive waveform to achieve the desired spectral output.

2.3. Constant-power amplifier implementation for surveillance sonars

Even for this subset of the power amplifier family there are numerous architectural combinations which offer particular advantages in their various applications. However, the basic ingredients of any amplifier are summed up in the functional block diagram of Fig. 2.1.

2.3.1 *The input stage*

The input stage comprises a rectifier, an input filter and a power regulator. Recent advances in power semiconductor technology (particularly power mosfets and the newer fast bipolar transistors) allow the power regulator, which is a form of switched mode power supply, to run at a frequency in the region of 100 kHz with good efficiency. This allows the size of associated inductors and capacitors in that stage and the input filter to be reduced. The power regulator will remove rectified mains ripple and ensure that the power throughput to the load is constant.

The function of power regulation may be accomplished in one stage as shown, or in two stages, by making additional use of the inverter drive to regulate current. The point of feedback is also variable. Current and voltage can be sampled at the output (impedance matching) point for high accuracy, and multiplied together to give a power-level feedback. It is also feasible to implement the power regulation solely by modulating the output bridge, though this may lead to a restriction in the dynamic range of the output power. Voltage and current can alternatively be sampled at the inverter input, and used to derive a constant-power control.

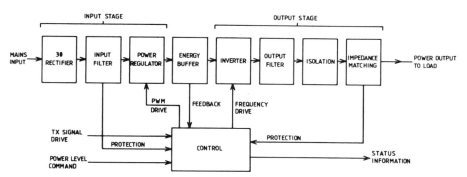

Fig. 2.1. *General block diagram of a constant power amplifer*

2.3.2 *Energy buffer*

This is a useful practical concept in power amplifier design. In general, the power to the amplifier load will vary instantaneously over each waveform output cycle. In particular, sonar transducers can be highly reactive, so that over portions of the cycle considerable energy is fed back into the amplifier from the load. Unless an energy buffer of suitable capacity is present, ultimately the energy variation is coupled back to the mains, which is usually desirable! In the case of power regulators which derive feedback from the energy buffer stage, the ripple of the measured quantity (voltage across a capacitor for the voltage fed output stage or current in an inductor for a current fed configuration) effectively limits the high-frequency gain and therefore the transient response of the system. Note that this constant is not normally present in switched-mode power supplies, where output-ripple requirements usually dictate a larger energy store. It is, therefore, this stage, particularly in the low-frequency designs, which often limits the compactness of the hardware. Another undesirable side effect of reducing the energy buffer is to raise the stress levels on the power switching devices, thereby effectively constraining their useful power throughput.

2.3.3 *Output stage*

The output stage will comprise an inverter, which is usually a bridge configuration, an optional output filter, an isolator, which could alternatively be associated with the power regulator, and impedance matching.

Sonar transducers are normally parallel tuned, so the load impedance tends to zero for frequencies well above or below the operating band. For a class-D output stage this immediately suggests a current-fed design, obviating the need for an output filter to limit harmonic current. However, in practice, the level of output harmonics (and switching noise) is of concern and a voltage-fed configuration will be adopted with a series L-C output filter. Note that the absence of an output filter in current-fed designs gives an inherent wideband capability.

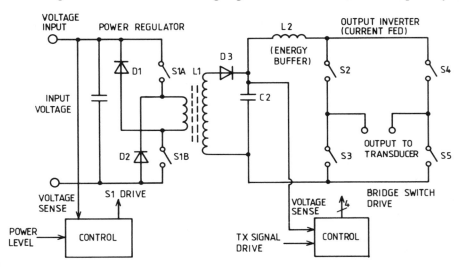

Fig. 2.2. *Feedforward control*

The mains-isolation and impedance-matching function for higher-frequency designs is implemented in a transformer at the inverter output. This offers the advantages of decoupling the specific impedance operating range of the load from the output stage, allowing the operating point of the inverter power-switching devices to be optimised. However, for lower-frequency amplifiers the output filter and transformer often moves to the input of the inverter, where it transfers power at the power-regulator switching frequency. The tuning inductor at the sonar transducer provides a further opportunity to impedance match. In some applications, such as helicopter dipping sonars, where transmission is required down a long cable at high voltage, the absence of the output transformer means that special design techniques must be employed to achieve high voltage switching in the output inverter.

The control system will typically implement protection on sensing either overcurrent in the output circuit or overvoltage (and undervoltage) at the mains input.

2.3.4 *Techniques of power regulation*

Types of constant-power regulator fall into two categories, either feedforward or feedback. Examples of each type are shown in Figs. 2.2 and 2.3.

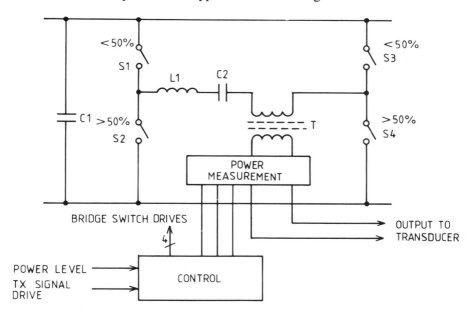

Fig. 2.3. Feedback control

2.3.4.1 *Feedforward regulation*

This is achieved by repetitively charging a reactive component to a defined level of energy and discharging that energy towards the output stage. Fig. 2.2. shows an implementation using an inductor, which also provides the isolation function. The configuration is recognisable as the familiar switched-mode power-supply "flyback" converter (1). During the charge period, S1A and S1B are turned on and current rises in L_1. When S1A and S1B are turned off, the energy in the inductor discharges into C_2. The power fed forward is given by:

$$P = \frac{(Vin \cdot t)^2 f}{2L_1}$$

where f is the operating frequency, and t is the conduction period for S1A and S1B.

The control system takes out variations in *Vin* by ensuring that *Vin.t* is a constant as required by the demanded power level. D_1 and D_2 limits the off voltage across S_1 to *Vin*. The Feedforward System has several advantages:

    i) Very fast transient response.
    ii) Unconditionally stable.

iii) Energy buffer size can be minimised, at the expense of higher stresses in S_1 and the output bridge.
iv) Simple control.

The stress in S1A and S1B can be lowered by pulse-width modulating the output bridge to take out reflected load impedance variations, allowing the voltage across C_2 to be stabilised at a high value. This allows a maximum duty cycle to be obtained in S1A and S1B.

Feedforward configurations in general have the disadvantage of higher peak-current stresses in the power regulator switches, which is particularly serious if "on-resistance" devices like power mosfets are employed.

2.3.4.2 Feedback control

In the feedback configuration of Fig. 2.3, the output voltage and current in T_1 is sensed. Power is obtained by multiplying the two values in a 4-quadrant multiplier and this is compared against a power-level reference. For a power output increasing above the reference level, the duty cycles of S_1 and S_3 are reduced relative to S_2 and S_4, thereby reducing the voltage (and hence power) delivered to the load. Note that the transient response is limited by the smoothing necessary to the power-feedback signal after multiplication.

In an alternative feedback control configuration, the voltage and current at the input of the inverter stage can be separately sampled. The voltage signal is fed back to control for constant voltage by regulating the power-converter stage inserted before the output bridge. The current, sampled in a series resistor, is fed forward to control the duty cycle in the output bridge, so as to give constant current. Thus both control loops work together to give constant power operation. The advantage of this mode of operation is that the "variable" dynamic range in the two control systems is the square root of that in a power feedback system, over a given power dynamic range. This allows accurate power control to be established in applications where a large variation in the amplitude of the output is required.

Acknowledgements

The author wishes to thank P. Busby, PMRU, for his contribution in compiling this paper. Also, G. Doye, PMRU, for his early efforts in generating enthusiasm for this work.

Reference

1. JANSSON, L. E., 'A Survey of Connector Circuits for Switched Mode Power Supplies'. Mullard Technical Note 24; TP1442/1 1975.

Chapter 3
Radar transmitters

N. S. Nicholls
Consultant to Marconi Radar Systems Ltd

3.1 Introduction

A radar transmitter may be regarded both as a power converter and as a signal synthesiser or amplifier. The latter viewpoint is of primary interest to this symposium. However, the performance attainable in this regard is usually constrained by conflicting considerations of power conversion performance, so this aspect will be examined first.

3.2 The transmitter as a power converter

3.2.1 *Aspects of power conversion performance.*
The most significant aspects of power conversion performance are listed below. The order is arbitrary, their relative importance depending on the application.

a. Output power
b. Efficiency
c. Size and weight
d. Reliability and availablity
e. Ruggedness
f. First cost
g. Support cost and maintainablity
h. Electromagnetic compatibility (e.m.c.)
j. Start up delay
k. Shelf life

Most radars employ pulsed transmission. The Duty Ratio (the fraction of time occupied by transmission) typically lies in the range 0.1% and 10%.
  C.W. Systems are sometimes used, the target echo being distinguished by its Doppler Shift, but they demand a transmission of great spectral purity.

16 Radar transmitters

3.2.2 Device types
Most radars radiate in the frequency range between 1 GHz and 100 GHz. Over much of this range the choice of power conversion technologies is the same. The important microwave amplifying and oscillating devices (the distinction between these modes of operation is incomplete) are:-

Travelling-wave tube (t.w.t.)	} Linear	
Klystron	} Beam	} Vacuum
Cross-field amplifiers (c.f.a.)	} Crossed	} devices
Magnetron	} field	
Transistors	} semiconductor	
Bulk effect devices	} devices	

3.2.3 Power range
Vacuum devices can be designed for much greater powers than semiconductor devices. The maximum power possible with all types of device falls rapidly with increasing frequency. The achievement of adequate power is not a serious problem when using valves, except at the highest frequencies. Vacuum devices are used at output powers of up to a few megawatts peak and a few tens of kilowatts average at frequencies up to 6 GHz.

At low powers (below 1 kW peak, 100 W average) crossed field tubes and semiconductor devices offer the best power conversion performance.

High powers may be generated by the use of a multiplicity of low power transmitters. A potential advantage is greatly improved reliability if failure of a fraction of the transmitters can be tolerated. At the lower frequencies, transistors are often applicable. Schemes of this type are attractive where a separate antenna is provided for each transmitter, thus both minimizing coupling between the separate units and also permitting electronic steering of the radiated beam by phase-shifting of the drive signals.

3.2.4 Efficiency
The natural efficiency of crossed-field tubes and transistors, 35% to 70% depending mainly on circuit losses, is superior to that of linear-beam devices particularly t.w.t's but the efficiency of the latter may be raised by depressing the collector potential. This is most straightforward in tubes with d.c. h.t. supply (gridded gun tubes) where any variation in circuit interception current (which may depend on frequency and drive power) may be absorbed during each pulse by reservoir capacitors and compensated during the interpulse interval.

The overall efficiency of the transmitter is reduced below that of the microwave devices by losses in converting the prime power into appropriate device supplies and also by power consumed in providing cooling and other services. An efficiency as high as 90% may be achieved in the provision of accurately controlled d.c. h.t. supplies. An efficiency of 80% is attainable in pulsed h.t.

supplies such as are required by magnetrons and some other types of device if limited flexibility in pulse duration and spacing are acceptable.

3.2.5 *Size and weight*

Crossed-field tubes are smaller in size and weight than linear-beam tubes. The advantage is most marked for magnetron oscillators at the lower frequencies (see Fig. 3.1). Much of the size advantage of c.f.a's is offset by their low gain,

Fig. 3.1. *Dual 650 Kw peak 550 w average power 3 GHz transmitters and receiver rack, showing coaxial magnetron with modular semiconductor modulator below*

which makes the driver stages comparable in size with the output stage. Semiconductor devices are very small but the low stage gain of transistor amplifiers leads to a significant overhead in r.f. circuit components. On the other hand, such amplifiers may be operated in Class C using d.c. supplies which eliminates all pulse modulation equipment (See Fig. 3.2).

3.2.6 *Cost*

The overall transmitter cost is usually several times greater than the cost of the microwave devices. Linear-beam tubes tend to make for the most expensive transmitters. Semiconductor transmitters are expected to benefit most in cost from manufacture in quantity.

18 Radar transmitters

3.2.7 *Electro-magnetic compatibility (e.m.c.)*
All radar transmitters tend to emit harmonics and noise, and spurious signals outside the signal bandwidth, and sometimes inter modulation products between certain of these. Crossed-field devices are usually rather worse than linear beam tubes in this respect. High power transmitters usually need a filter in the transmitter output waveguide.

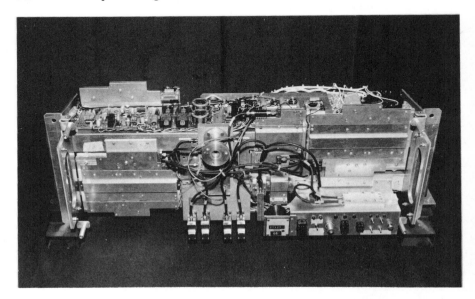

Fig. 3.2. *A 3.3 Kw peak 150 w average power 1.25 GHz transmitter with 24 dB gain using bipolar transistors. All power supplies are included*

3.2.8 *Reliability and maintainability*
Transmitters for the higher powers and the higher frequencies tend to fall short of desirable reliability. Tube transmitters employing cathode modulation have proved more reliable than those using d.c. h.t. (gridded gun) linear-beam tubes. Semiconductor devices may prove more reliable than tubes, though they are more vulnerable to abnormal conditions.

Linear-beam tubes sometimes cause maintainability problems at the lower frequencies, owing to their size.

3.2.9 *Ruggedness*
Semiconductor devices are much more rugged than vacuum tubes. Among the latter, ruggedness tends to be inversely related to size, but otherwise little affected by type.

3.2.10 *Start-up delay*
Linear-beam tubes typically require several minutes for cathode heating. Some

c.f.a.'s have cold cathodes and are instantly ready; however, owing to their low gain a transmitter employing them will usually contain t.w.t driver stage. Small hot-cathode crossed-field devices can be made to start in a fraction of a second. Only these and semiconductor transmitters offer a rapid start capability.

3.2.11 *Shelf life.*
Vacuum microwave devices can be built to have shelf lives of a decade or more, but only with moderate reliablity and at the expense of extra care in manufacture. There should be no problem with semiconductor devices.

3.2.12 *Conclusions on the power conversion aspect*
It is evident that the above factors are not free from mutual conflict, but collectively they will tend to be in opposition to signal quality; also that the application will often exert a strong pressure in favour of adopting a particular power conversion technology. The possibility of relaxing the transmitter waveform specification by the application of clever signal processing techniques may therefore be important.

3.3 The transmitter as a signal synthesiser or signal amplifier

3.3.1 *Types of defect in the transmitted signal*
Defects in the output signal of a transmitter which may degrade radar performance are:

1. Within-pulse repetitive modulation
2. Frequency distortion
3. Low-frequency (pulse-to-pulse) modulation
4. Wide band noise

The allocation of transmitter shortcomings between some of the above categories involves fine distinctions, so we will preface the comparison of available devices with a fuller definition of each category of defect.

The argument will be developed partly in the frequency domain. There a truly repetitive train of pulses appears as a group of lines spaced by the pulse repetition frequency (p.r.f.) The envelope (amplitude and phase) of the lines will determine the pulse shape.

3.3.2 *Within-pulse repetitive modulation*
If the transmitter were instructed to produce a train of identical pulses with fixed centre frequency, then the average (over a large number of pulses) departure of the output signal from the ideal in amplitude and phase as a function of time, but excluding that part attributed to frequency distortion (see 3.3.3 below) will be called within-pulse modulation.

It is not generally useful to distinguish 'amplitude distortion' from within-pulse modulation because it is normal to operate microwave power devices in a near-saturated condition – this means that they exhibit the characteristics of a limiter. This is done primarily with the object of maximising power output or minimising dissipation. Thus the 'ideal' output waveforms for radar transmitters almost invariably have a rectangular envelope.

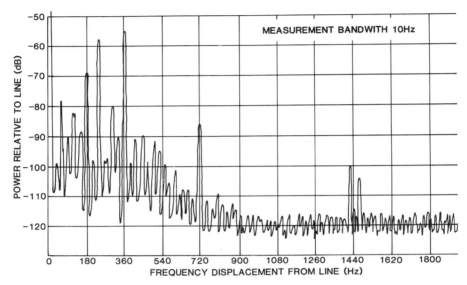

Fig. 3.3. *A spectrogram of interline noise for a t.w.t transmitter, showing modulation by harmonics of the 60 Hz prime power*

Within-pulse modulation is most severe in cathode-pulsed high-gain linear-beam tubes and is predominantly phase modulation on account of their large phase length. Hence their delay is very sensitive to h.t. voltage. By definition it affects only the amplitudes and phases of the spectral lines. It is particularly harmful to range resolution in radars using pulse-compression.

Within-pulse frequency modulation of magnetron oscillators may limit the maximum pulse length at which they can be used.

3.3.3 *Frequency distortion*
If the mean error of amplitude and phase at each instant, as described in 3.3.2 shows some correlation with the centre frequency of the signal then that part will be ascribed to 'frequency distortion' rather than to within-pulse modulation. The distinction is mainly of interest to transmitter engineers for the purpose of controlling the sources of signal distortion. For a fixed-frequency, fixed-waveform radar, frequency distortion would not be distinguishable in its effect

from within-pulse modulation; for a randomly frequency-agile radar, the effect may not be distinguishable in its effect from noise modulation. Frequency distortion sets a limit to the bandwidth over which stagger-tuned klystrons can be used with wide band signals.

3.3.4 *Low frequency (pulse to pulse) modulation*
Modulation at frequencies which are much less than the reciprocal of the pulse duration arises mainly from sources outside the tube and will have negligible effect on the shape of individual pulses. In the frequency domain, side-bands will be generated about each line. The consequent infilling of the space between the lines, which will exhibit high correlation between frequencies separated by one p.r.f. will be detrimental in systems relying on Doppler shift to detect targets in heavy clutter.

Modulation by harmonics of the power supply frequency and by low frequency noise is frequently encountered with high-gain linear-beam amplifiers (See Fig. 3.3).

3.3.5 *Wide-band noise*
Noise of bandwidth large compared with the reciprocal of the pulse duration is generated by microwave devices during operation. Because the noise is pulse-modulated, it will exhibit a similar correlation structure in the frequency domain to the low-frequency modulation side-bands. It is sufficiently strong in some c.f.a's and bulk-effect semiconductor devices to affect target detection in severe clutter.

3.4 The effects of defects of transmitter signal quality upon the performance attainable in radar systems

Small deficiences of signal quality from the transmitter of a radar may affect system discrimination in range and velocity. The extent to which such deficiencies may be compensated by clever processing of the receiver signal is outside the proper scope of this paper. It must suffice for us to say that, neglecting the complicating effects of receiver amplitude distortion the process of reception may be regarded as an examination of the cross corelation between the received signal and another somewhat similar signal displaced in time and frequency to take account of target range and velocity.

The sensitivity of the radar would be at a maximum if this latter signal, apart from the displacements, were a replica of the transmitter signal. However, particular aspects of the discrimination may be improved, at some cost in sensitivity, by departing from a replication of the transmitted signal in an appropriate way. The possibility also exists of changing this receiver waveform on a pulse-to-pulse basis in order to partially compensate for transmitter fluctuations.

3.5 The measurement of transmitter signal quality

Radar system discrimination requirements are often severe, so that the relevant imperfections of the transmitter are impossible to monitor by direct observation of waveforms. Instruments embodying signal processing techniques are required to provide data useful for diagnosis and adjustment.

Chapter 4
Receiver array technology for sonar

Iain Watson
ARE Portland

4.1 The problem

In sonar we are concerned with the detection of targets against a background of unwanted signals. These unwanted signals may be noise in several forms, or reverberation. The task of the receiving system designer is to ensure an optimal system response to the wanted signal and minimal response to the unwanted signal. The receive array can play an important part in this process.

Sonar suffers from noise problems due to its use of sound waves whose influence on the receiving sensor may be indistinguishable from that due to other mechanical sources. Discrimination against vibration for example may be best achieved in the sound sensor itself rather than in the signal processing. This paper will be concerned with the very early stages in the sonar detection process and I will try, as far as possible, to neglect the signal processing. Other speakers will cover that topic far better than I could.

This paper will also neglect the problem of reverberation in active systems. The essential key to reduced reverberation input is the use of high resolution and optimum frequency, and this depends so fundamentally on the role of the system as to make general comments unhelpful.

We are therefore left to consider the rejection of unwanted input noise or *the signal conditioning of the acoustic problem.*

4.2 The sonar array

Consider a typical sonar array. It comprises a number of hydrophones, a signal conditioning baffle and a mounting system. In large ship or submarine mounted arrays it is usual to provide a sonar dome to stand the flow away from the sensor surface. We shall consider very briefly the part that each of these components can play in the rejection of the various noise sources.

4.3 The hydrophone

The hydrophone is an electrochemical device that measures pressure. Both the pressure field and the electrical components of the hydrophone will be subject to thermal noise. A simple analysis of the hydrophone response yields the not surprising result that a lossless 100% efficient hydrophone produces a thermally induced electrical noise output equivalent to the thermally induced pressure field around it. A real hydrophone has a quite different behaviour and it is necessary to consider both the internal losses and the mechanical resonances of the device. Quite simple design criteria can be arrived at for frequencies up to about 40 kHz, where the thermal noise is much lower than the lowest observed sea state noise levels. Above this frequency the situation becomes much more difficult and both high efficiency and sensitivity are required. Careful matching to the preamplifier is required for optimal noise performance.

It is also desirable that the hydrophone have a low sensitivity to acceleration. An upper limit for acceleration rejection can be deduced from the plane wave pressure implied by the acceleration of the baffle. Highly satisfactory rejection of acceleration response has been achieved by both suitable design of the mounting system and the use of acceleration cancelling sensors.

4.4 The baffle

The purpose of the array baffle is to provide both optimum pressure response and maximal noise rejection. Analysis of the baffle rapidly involves the use of complex mathematics but the basic physical requirement can be simply stated as the achievement of a high transmission loss at the frequencies of interest. This has been attempted by the use of either high (eg. steel) or low (eg. polyurethane foam) impedance materials. These do all that can be desired provided that they actually satisfy the soft or hard requirement.

In practical materials however the impedance mismatch actually achieved is finite, and this severely restricts the frequency range over which an adequate baffle performance can be obtained. Some typical response curves have been calculated using an infinite flat plate model of the baffle. These demonstrate the difficulties inherent in the two types of material.

4.5 Flow noise and domes

Starting from Lighthill's model of the turbulent boundary layer and making some fairly sweeping assumptions about the nature of the boundary it is possible to arrive at a model of the surface pressure spectrum. This is analysed in terms of wavenumbers which represent both acoustic and non-acoustic components of the pressure field. There are two peaks in this spectrum, one at near sonic velocity representing an acoustic signal at near grazing incidence, the other at a typical flow velocity corresponding to the convection of small non-acoustic pressure fluctuations. These non-acoustic components are non-radiating and decay away from the boundary. Although they represent a fluctuating pressure field they do not satisfy the wave equation and are therefore refered to as *pseudo sound*.

A large dome-array stand off significantly reduces these contributions and hence the associated flow field. Another means of reduction is to "average" these slowly propagating disturbances by using hydrophones of finite size.

References

1. DOWLING, A. P. and FFOWCS-WILLIAMS, J. E., 1983, 'Sound and Sources of Sound', John Wiley.
2. ROSS, D., 1976, 'Mechanics of Underwater Noise', Pergamon Press.

Chapter 5

New underwater acoustic sensors

M. L. Henning and R. J. Langston
Plessey Marine Ltd U.K.

5.1 Introduction

The renaissance of optical technology as a result of the development of coherent optics and optical fibres has led to a new class of transducers based upon the microphonic nature of optical fibres. Prominent among these sensors is the optical fibre hydrophone. Meanwhile, less radical approaches to the drawbacks of conventional piezo-electric hydrophones, involving a better understanding of materials behaviour, have resulted in the development of a variety of composite materials formulations. New fabrication techniques have provided the means for new configurations using conventional materials, thus reducing constraints in the design of transducers. Recent developments in these areas are described in this chapter together with some discussion on techniques and applications.

5.2 The optical fibre hydrophone

5.2.1. Basic mechanisms

The microphonic properties of optical fibres have been used for the detection of sound for a number of years. [Cole *et al.* (1)]. Both amplitude modulation and phase modulation of the light travelling within the fibre have been used. In principle, the latter mechanism is by far the more sensitive, and it is this with which we are concerned. The effect of pressure upon the optical path distance in a fibre is to change the physical length of the fibre, and its refractive index. This is manifested by a change in the phase of the light leaving the fibre. Hence,

$$\delta\phi_n = L \cdot \delta\beta_n + \beta_n \cdot \delta L$$

where L is the fibre length and β_n is the propagation constant for the nth mode in the fibre.

In the case of hydrostatic pressure upon a bare fibre, it is easy to show that the change of phase $\delta\phi$ with pressure δP is given by

$$\frac{\delta\phi}{L\delta P} = \frac{\kappa_0 \mu (1 - 2\sigma)}{E} \left[1 - \frac{\mu^2}{2} (2p_{12} + p_{11}) \right]$$

where, σ is the Poisson's ratio of the fibre, E is Young's modulus, μ is the refractive index, p_{12}, p_{11} are the photoelastic strain coefficients and κ_0 is the free space propagation constant. For a typical fibre,

$$\frac{\delta\phi}{L\delta P} \simeq - 210 \,\text{dB re 1 rad } |\mu\text{Pa}|\text{m}$$

In the case of a jacketed fibre, the equivalent sensitivity is given by Hocker (2), and is now also a function of the radii of the jacket and fibre and the elastic constants of the jacket as well as the fibre. It is important to note that the sensitivity of such a packaged fibre depends upon the mode of constraint of the fibre, and, in general, the hydrostatic mode treated here is the least sensitive. The variation of sensitivity with jacket material is illustrated in Figure 5.1. It can be seen that the use of a jacket gives, typically, 20 to 40 dB more sensitivity than the bare fibre.

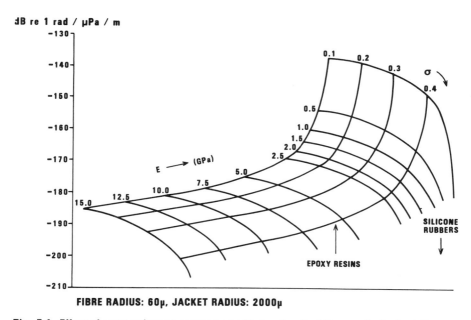

Fig. 5.1. *Effect of encapsulant on pressure sensitivity of optical fibre under hydrostatic conditions*

5.2.2 Detection techniques

As in other optical systems, interferometric techniques are used to detect phase or path length changes; interferometers may be either base band (homodyne), or heterodyne systems using an intermediate frequency carrier. The former are more sensitive, but the latter are generally more stable. A typical configuration is that of the Mach Zehnder, or ring interferometer, in which the intermediate frequency shift is provided by a Bragg cell in the reference arm [Hall, (3)]. Light arriving from the sensor arm, with its acoustic phase modulation, is mixed at the photodiode with the up- (or down-) shifted reference light. The output of the diode is a phase modulated carrier at the intermediate frequency. The signal can be extracted from this carrier by conventional techniques.

5.2.3 *Practical systems*

The linear nature of the fibre sensor holds forth the possibility of very large scale underwater sensors. Current optical-fibre attenuation figures of less than 1 dB/km, and projected figures of less than 0.1 dB/km encourage this speculation. In addition, the non-electrical nature of the transduction mechanism raises the question of whether a totally electrically passive underwater system can be achieved, with its attendant advantages. In principle, the homodyning Mach Zehnder interferometer can be used in this way, but it is unstable; alternatively the sensing part of a heterodyne system could be put underwater, and the reference kept dry; however, the leads to the sensing element in this scheme are acoustically sensitive.

One practical way of achieving a stable passive remote optical sensor is the differential delay heterodyning interferometer [Henning *et al.* (4)]. In this configuration, optical pulses of alternating frequencies are sent to the sensor, where the sensor fibre length is arranged to cause an optical delay of one pulse length. The differential delay between sensor and reference in the interferometer causes two pulse trains to return to the photodetector via the single return fibre and displaced by one pulse length. At any given time there are two optical frequencies present, thereby providing the conditions for heterodyning. Environmental disturbances on the leads (which are of long period compared with the optical pulses) are common to both pulse trains, and are thus rejected in the mixing process.

A variation upon this technique, and one which forms the basis of a practical hydrophone is the reflectometric sensor [Dakin *et al.* (5)] (Fig. 5.2).

In this configuration, the sensor consists of a downlead, a semi reflecting fibre splice, a sensor fibre, and another reflecting, or semi reflecting surface at the end of the sensor. An array of sensors can be achieved simply by replacing the end reflector by another semi reflecting splice, followed by another, and so on.

As in the previous example, the pulse length is made equal to the time of transit of the light in the sensing fibre. In the simple single sensor, the train of pulses of alternating frequencies can be used as before, or gaps can be left between pairs of pulses, to allow for multiple reflections. As before, environmental effects on the leads are cancelled out. The output from an array of such sensors is a time-division multiplexed signal, which can be used for electrical beamforming, after decoding into the individual acoustic outputs. Laboratory prototypes of these sensors can provide typical system sensitivities of better than 20 dB below sea-state-zero noise levels at 1 kHz, with a sensor fibre length of about 100 m.

5.2.4. Applications

The potential advantages of optical hydrophone arrays appear to be in areas currently served only with difficulty by conventional transducers. In particular, very large scale arrays, situated remotely from receiving stations, and requiring very high reliability, could be realised using optical fibres. Large two-dimensional arrays of transducers with aperture shading, to operate at low frequency can, in principle, be economically realised using this technology. The high bandwidth attainable with optical fibre cables, coupled with the essentially non-electrical nature of the fibre hydrophone provides the possibility of non-electrical towed hydrophone arrays.

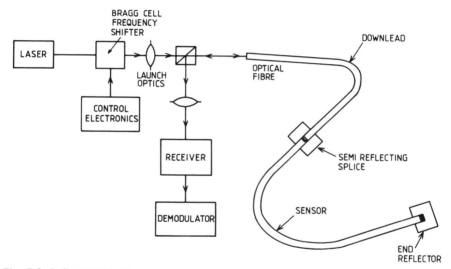

Fig. 5.2. Reflectometric fibre sensor

The usual configuration of optical fibre hydrophones is with the fibre jacketed to an overall diameter of about 1 mm and wound into an appropriate shape, or with the fibre wholly encapsulated within a water compatible material, such as an epoxy resin or rubber. Either of these configurations can provide improvements in immunity to acceleration forces, compared with high density piezo-electric materials; in addition, such sensors will have a low acoustic-impedance mismatch with water, and therefore low reflectivity.

5.2.5 Conclusions

Practical first-generation fibre optic hydrophones are close to realisation, using pulse techniques for achieving non electrical undersea operation. Techniques are

30   New underwater acoustic sensors

available for constructing and deploying arrays of such sensors. In the longer term, the practicality of large remote arrays (phased or unphased) will be demonstrated, and it is in this area that the first applications are expected.

5.3. Composite materials

5.3.1 *Rationale*

The trend in military sonars is to higher detection ranges with a consequential lowering of the operational frequency to minimize the losses due to acoustic absorption in the sea. This concept in many cases leads to individual sensors effectively being subjected to an hydrostatic pressure field. In many other cases, the mechanical construction of arrays could be made simpler and less costly if the transducers could be completely exposed to the hydrostatic pressure field.

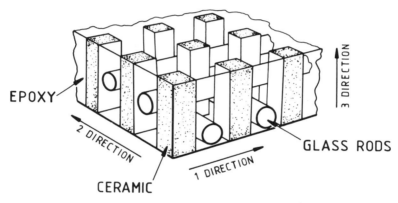

Fig. 5.3. *Typical 3-phase composite piezo-electric material*

Unfortunately, the hydrostatic pressure sensitivity of most currently available ceramics is much less than that of the thickness mode (see Fig. 5.3–3 direction), due to the inherently antiphase coupling to the mutually orthogonal planes (directions 1 and 2). This low sensitivity is conventionally overcome by sheilding the ceramic from acoustic radiation in the 1 and 2 directions, with complex transducer housings.

5.3.2 *Principles*

The concept of composite transducers (Shorrocks *et al.* (6)) arises from the desire to depart from the rather rigid constraints put on transducer design by the use of conventional piezo-electric ceramics. The aim is to produce a lightweight piezo-electric 'ceramic' with enhanced hydrostatic performance and without the consequent requirement for complex housings. A schematic explanatory diagram showing the structure of a typical three-phase composite is shown in Fig. 5.3. The three phases relate to the number of different materials embodied in the main composite structure; in this case the phases are the ceramic, glass rods and encapsulant.

The mechanism by which the hydrostatic sensitivity is enhanced is complex, consisting of several effects, and has not yet been fully resolved. A simple insight is that the stiffness or resistance to motion in the 1 and 2 directions has been increased by the addition of the appropriate non ceramic materials between the ceramic pillars.

A typical example of development work involves conventional PZT-5H ceramic cut into 0.6 mm square rods with either pyrex or soda glass rods of 0.25 mm diameter. The encapsulants are usually epoxy resins of varying mechanical properties. The electrodes have mainly been of the conducting paint type although some composites have been produced with thin copper-plate electrodes.

5.3.3 Performance

A figure of merit for the hydrostatic sensitivity performance of the various composite fabrications is given by the product of voltage and charge coefficients of the composite (the $g_h^1 d_h^1$ product) and is a measure of the total energy generated per unit volume. The results on the composites tested to date show that for a given set of materials there exists a ceramic volume fraction (the ratio of the ceramic volume to the total composite volume) at which the value of $g_h^1 d_h^1$ is an optimum, and which approaches in some instances the $g_{33} d_{33}$ of the basic ceramic. Composites of various sizes have been constructed, maintaining a constant ratio of the constituent phases, to test the scalability of the devices. Results show that the hydrostatic performance of the composite is unaltered by making a large area device. An interesting point to note is that air bubbles trapped in the encapsulant can give a large increase in the $g_h^1 d_h^1$ product although why this should occur is not yet clear.

Although the $g_h^1 d_h^1$ product is a useful yardstick for research and development purposes it is not always the most useful parameter for the transducer designer, who will usually be tasked with producing a voltage or charge generator rather than an energy efficient device. Fortunately either of the two hydrostatic coefficients can be readily optimised for this purpose.

Testing of the composites as acoustic transmitters has also been performed and they have been shown to have useful transmitting sensitivities. The thickness coupling coefficient (K_t) is generally high and shows a marked tendency to increase with increasing ceramic volume fraction as does the frequency constant. The frequency constant is also a function of the stiffness of the encapsulant. Hence the designer in principle can realise a large-bandwidth device by locally changing the ceramic volume fraction or the mechanical properties of the encapsulant.

The noise performance of a system incorporating composite transducers generally will improve, since the inherent device noise is negligible. The extent of the improvement will depend upon the origin of the other noise sources in the system, and the configuration of the device as either a charge or voltage generator.

5.3.4 Applications

The advantages of three-phase composite materials for use in acoustic transducers can be summarized as follows:

(a) High hydrostatic pressure sensitivity can easily be engineered.
(b) Good design flexibility for either charge or voltage sensitivity is available.
(c) The weight and dielectric constant can easily be tailored to specific requirements.
(d) The 'matrix' structure of the ceramic elements lends itself to mechanical or electrical shading of the composite structure.
(e) Array construction is lightweight and simple.
(f) Devices have a useful transmit capability.
(g) It is possible to provide a tuning capability within the composite structure to produce high-bandwidth devices.

Areas of application which would benefit from this technology include large-area low-frequency passive arrays, where the enhanced sensitivity allows active areas of small size, weight and cost. Other applications include high-frequency wide-bandwidth active arrays and single transducers with aperture shading.

5.4 Multilayer transducers

A major problem in many transducer designs is the interface between the transducer and the electronic system, where impedance matching becomes an important consideration. One configuration of elements in a transducer stack which provides considerable flexibility in this respect is shown in Figure 5.4.

The multilayer "ceramic", consists in this case, of three layers with the poling directions reversed, and electrically connected in parallel. As a result, the layers are mechanically in series and the mechanical behaviour of the ensemble is the same as that of a conventional ceramic transmitting transducer of the same size and shape. However, analysis of the electrical characteristics shows that the transducer impedance and voltage drive can easily be chosen to suit the desired design objectives. For example, for a given acoustic output,

$$\text{The impedance } Z_n = \frac{Z_1}{n^2}$$

$$\text{The drive voltage } V_n = \frac{V_1}{n}$$

where subscript (n) refers to the n layered device, and (1) refers to a single element of the same size and shape.

It is apparent that one advantage of this configuration is that the problem of designing appropriate matching transformers is eased.

In the case of a receiving transducer, the multilayer sensor may be reduced in face area by a factor n^2 compared with a conventional transducer of the same capacitance. Such a sensor has charge and voltage sensitivities increased by a factor n. The improvement in noise performance of a system with these devices will depend upon the origin of the noise; if it is system-related, an improvement in signal-to-noise ratio of n^2 could be expected in the above example.

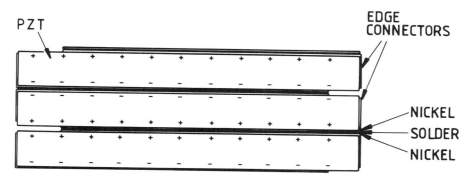

Fig. 5.4. *Schematic of multilayer transducer*

It can be seen that this arrangement of stacked reverse-poled layers of piezo electric devices, connected in parallel, leads to a broadening of the options available to the transducer designer which should ease the acoustical and electrical design of sonar systems.

5.5 Acknowledgements

This chapter is based upon work carried out at various Plessey Research Establishments, and the authors especially wish to acknowledge the contributions of the following:

Mr. S. Thornton   (Plessey Marine Research Unit)
Dr. J. Dakin   (Plessey Electronic Systems Research)
Mr. C. Wade   (Plessey Electronic Systems Research)
Dr. R. Whatmore   (Plessey Research)
Mr. A. G. Munns   (Plessey Research)

5.6 References

1. COLE, J. H., JOHNSON, R. L., BHUTA, P. G., 1977, *J. Ac. Soc. Am.*, **62**, 1136–1138.
2. HOCKER, G. B., 1979, *App. Optics* **18**, 3679–3683.
3. HALL, T. J., *PROC. IEE(H)* **127**, 193–200.

4. HENNING, M. L., THORNTON, S. W., CARPENTER, R., STEWART, W. J., DAKIN, J. P., WADE, C. W., 1983, 'Optical Fibre Hydrophones with Downlead Insensitivity', IEE Conference Proceedings 'Optical Fibre Sensors', London, England.
5. DAKIN, J. P., WADE, C. W., HENNING, M. L., 1984, *Elect. Lett.* **20**, 53–54.
6. SHORROCKS, N. M., BROWN, M. E., WHATMORE, R. W., AINGER, F. W., 1982, 'Piezoelectric Composites for Underwater Transducers', Conference Proceedings, 1982 European Ferroelectrics Meeting, Malagar, Spain.

Chapter 6
Diversity techniques in communications receivers

J. D. Parsons
The University of Liverpool, UK

6.1 Introduction

In all except very simple transmission conditions, radio communication links are subjected to conditions in which energy can travel from the transmitter to the receiver via more than one path. This "multipath" situation arises in different ways depending upon the application, for example in MF and HF skywave transmissions it arises because of reflections from two or more ionospheric layers, or from two-hop propagation. In mobile radio it arises because of reflection and scattering from buildings, trees and other obstacles along the path.

Radio waves therefore arrive at the receiver from different directions, they have different time delays and they combine vectorially at the receiver to give a resultant signal which can be large or small depending upon whether the incoming waves combine constructively or destructively. An HF receiver at one location may experience a signal strength several tens of dB different from a similar receiver located only a short distance away. Even at one location however, the signal strength is not constant because as the ionospheric layers move with respect to the transmitter and receiver the phase relationship between the various incoming waves changes and the signal is said to be subject to fading. It is worth noting that whenever relative motion exists there is a Doppler shift in the received signal, this being a manifestation in the frequency domain of the envelope fading in the time domain.

In the mobile radio case, fading and Doppler shifts arise as a result of motion of the receiver through what is essentially a spatially-varying field. However, whatever the mechanism, the effect of multipath propagation is to produce a received signal with an amplitude that changes quite substantially with time. The temporal scale depends upon the specific case; in HF transmission ionospheric motion causes a fade every few seconds, whilst in VHF and UHF mobile radio, a vehicle moving at 50 km/hr can pass through several fades in a second.

36 Diversity techniques in communications receivers

6.2 Envelope properties

The different time delays mean that the incoming waves have different phases, and the resultant is therefore the sum of a number of sinewaves of random phase. This is a well-known mathematical problem which has received attention in the literature.

The resultant r can be written as

$$r \exp(j\phi) = \sum_{k=1}^{N} r_k \exp(j\phi_k)$$

Generally, r and ϕ can only be described in statistical terms and as $N \to \infty$ so

$$p_r(r) = \frac{r}{\sigma^2} \exp(-r^2/2\sigma^2) \qquad (1)$$

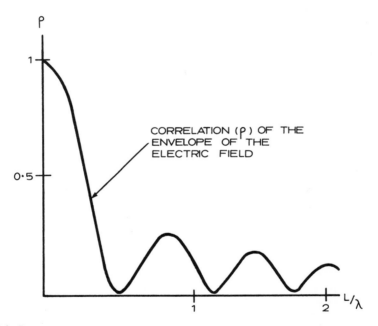

Fig. 6.1. *Envelope correlation function for an isotropically scattered field*

This is the RAYLEIGH distribution with a mean square value of $2\sigma^2$. At the same time it can be shown that the phase angle ϕ is UNIFORMLY distributed in the interval $-\pi < \phi < +\pi$, i.e.

$$p_\phi(\phi) = \frac{1}{2\pi} \qquad (2)$$

The Rayleigh distribution therefore describes the first-order statistics of the envelope. The probability that the envelope is below any specified value, say R is

$$P_r(R) = \text{prob}\,[r < R] = \int_0^R p_r(r)\,dr$$
$$= 1 - \exp(-R^2/2\sigma^2) \qquad (3)$$

From the cumulative distribution $P_r(R)$ we can find the overall fraction of time for which the signal envelope is less than any specific value. If we wish to know how this time is made up, then we need more detailed information such as the number of times the signal value crosses a given level, and on average how long it remains below that level.

One further property of the envelope that is of interest is the auto correlation function – it is relevant to the implementation of space diversity systems which will be discussed later. In isotropically scattered fields it can be shown that the spatial auto-correlation function of the signal envelope is given by

$$\rho_r(l) = J_0^2(\beta l) \qquad (4)$$

where l = distance, $\beta = 2\pi/\lambda$ and $J_0(.)$ is the zero-order Bessel function of the first kind. A plot of this function is shown in Fig. 6.1. It is worth noting that if the incoming waves are not isotropically scattered the decorrelation distances will be greater.

6.3 Bit error rate (BER) performance in fading conditions

Theoretical analyses of the error perfomance of data communication systems are usually based on the assumptions of slow, non-selective Rayleigh fading and multiplicative noise which varies slowly with respect to the signal period. These assumptions imply that the medium does not introduce any pulse lengthening so that there is no intersymbol interference, and that during any given symbol period the noise can be regarded as effectively constant. Of course, intersymbol interference can also arise due to the receiver filter, and radio frequency errors can be an additional source of distortion-related errors, but these are normally considered as negligible. Fortunately, in many types of radio channel the time delay spread is small compared with the bit rate, so whilst realising that these assumptions may not accurately represent the true situation, their acceptance is nevertheless justified for the purpose of analysis. On this basis we can calculate the probability of error in a fading channel by adapting the results for the probability of error for a steady signal in Gaussian noise. In the fading case these formulae can be taken to describe the probability of error for an individual pulse, conditional on the value of the multiplicative noise over that pulse, and hence they describe the conditional probability of detection errors caused by the additive stationary Gaussian noise. To determine the average performance over fading then requires averaging this conditional probability over the ensemble of possible values of the multiplicative factor, i.e. over the Rayleigh distribution for the signal. For the most common date transmission formats, the probability of error in Gaussian noise can be shown to be

Ideal coherent PSK: $\quad P_e = \tfrac{1}{2} \operatorname{erfc}(\sqrt{\gamma})$ (5)

Coherent FSK: $\quad P_e = \tfrac{1}{2} \operatorname{erf}(\sqrt{\gamma/2})$ (6)

DPSK: $\quad P_e = \tfrac{1}{2} \exp(-\gamma)$ (7)

Noncoherent FSK: $\quad P_e = \tfrac{1}{2} \exp(-\gamma/2)$ (8)

where γ is the carrier-to-noise ratio (CNR).

$$\text{CNR} = \frac{\text{signal energy per bit}}{\text{noise spectral density}}$$

$$= \frac{\text{signal power}}{\text{noise power in the bit-rate bandwidth}}$$

When the signal is subject to fading, the CNR will vary with time due to the changes in the signal level. We know that the signal envelope is Rayleigh-distributed but it is convenient, since we are interested in CNR to change the variable and find a probability density function for γ.

Consider a Rayleigh distributed signal with an envelope r described by (1) in the presence of additive Gaussian noise of mean power N. In these terms we can write

$$\gamma = \frac{r^2}{2N}$$

and we can define

$$\Gamma = \bar{\gamma} = \frac{\text{mean signal power}}{\text{mean noise power}} = \frac{\sigma^2}{N}$$

Simple manipulation gives

$$p(\gamma) = \frac{1}{\Gamma} \exp(-\gamma/\Gamma)$$ (9)

The probability that γ is less than some specified values γ_s is then

$$P(\gamma_s) = \operatorname{prob}[\gamma < \gamma_s] = \frac{1}{\Gamma} \int_0^{\gamma_s} \exp(-\gamma/\Gamma)\, d\gamma$$

$$= 1 - \exp(-\gamma_s/\Gamma)$$ (10)

6.3.1 Non-coherent frequency shift keying (FSK)

To find the error rate with non-coherent FSK we need to return to (8) which gives the BER for one value of γ and integrate it over all possible values of γ, *weighing* the integral by the probability density function (pfd) of γ, thus

$$P_e = \int_0^\infty \frac{1}{2} \exp(-\gamma/2) \frac{1}{\Gamma} \exp(-\gamma/\Gamma) \, d\gamma$$

$$= \frac{1}{2\Gamma} \int_0^\infty \exp\left[-\gamma\left(\frac{1}{2} + \frac{1}{\Gamma}\right)\right] d\gamma$$

$$= \frac{1}{2 + \Gamma} \tag{11}$$

6.3.2 Differential phase-shift keying (DPSK)
For DPSK we have

$$P_e = \frac{1}{2} \int_0^\infty \exp\left[\gamma - \left(1 + \frac{1}{\Gamma}\right)\right] d\gamma$$

$$= \frac{1}{2(1 + \Gamma)} \tag{12}$$

6.3.3 Coherent systems
For ideal PSK and coherent FSK the expressions for P_e involve the integral of an error function which is somewhat more difficult. For example in ideal PSK the required integral is

$$P_e = \frac{1}{2} \int_0^\infty \operatorname{erfc}(\sqrt{\gamma}) \frac{1}{\Gamma} \exp(-\gamma/\Gamma) \, d\gamma$$

This is not straighforward, but the result is

$$P_e = \frac{1}{2}\left[1 - \frac{1}{\sqrt{1 + 1/\Gamma}}\right] \tag{13}$$

and for coherent FSK

$$P_e = \frac{1}{2}\left[1 - \frac{1}{\sqrt{1 + 2/\Gamma}}\right] \tag{14}$$

6.3.4 System performance
Fig. 6.2 shows how the BER varies as a function of Γ, the CNR, for various communication systems. The severe degradation due to fading is obvious; for a single binary channel subjected to slow, non-selective Rayleigh fading, to maintain an error rate of 1 in 10^3 we require an increase of 20 dB in the mean CNR compared with the required CNR in the absence of fading. An additional 10 dB in CNR is required to reduce the BER by a further order of magnitude. The asymtotic perfomance in Rayleigh fading can be obtained from (11) to (14) and shows that at large CNR the error rate for all these systems is exactly inversely

proportional to the mean CNR as opposed to the exponential change in the steady signal case. This indicates the nature of the additional design burden imposed upon systems which are required to operate in fading environments and the resulting importance of techniques (such as diversity) for overcoming the problem.

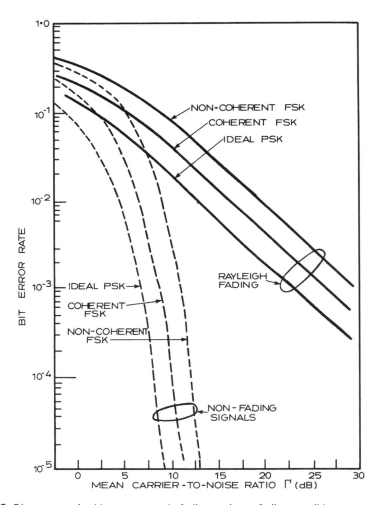

Fig. 6.2. Bit error rate for binary systems in fading and non-fading conditions

6.4 Diversity reception

The basic idea underlying diversity reception is that if two or more independent samples of a random process are taken, then these samples will fade in an uncorrelated manner. It follows that the probability of all the samples being simultaneously below a given level is very much less than the probability of any individual sample being below that level. In fact the probability of M samples all being below a certain level is p^M where p is the probability that a single sample is below the level, so it can be seen that a signal composed of a suitable combination of the various samples will have much less severe fading properties than any individual sample alone.

6.4.1 Basic diversity methods

Having obtained the necessary samples, we now come to another question which is how to process these signals to obtain the best results. There are various possibilities, but we must bear in mind that what is "best" really amounts to deciding what method gives the optimum improvement taking into account the complexity and cost involved.

For analogue communication systems the possibilities reduce to methods which can be broadly classified as "linear" combiners. In linear diversity combining the various signal inputs are individually weighted and then added together. If addition takes place after detection the system is known as a "post-detection combiner" and if before detection as a "pre-detection combiner". In the latter case it is necessary to provide a method of co-phasing the signals before addition. Assuming that any necessary processing of this kind has been done, we can express the output of a linear combiner consisting of M branches as

$$s(t) = \sum_{k=1}^{M} a_k s_k(t) \tag{15}$$

where $s_k(t)$ is the envelope of the k-th signal to which a weight a_k is applied.

The analysis of combiners is usually carried out in terms of signal-to-noise ratios (SNR), with the following assumptions

(a) the noise in each branch is independent of the signal, and is additive;
(b) the signals are locally coherent implying that, although their amplitudes change due to fading, the fading rate is much slower than the lowest modulation frequency present in the signal;
(c) the noise components are locally incoherent and have zero means, with a constant local mean-square (constant noise power);
(d) the local mean-square values of the signals are statistically independent.

Depending on the choice of the a_k's, different realisations and perfomance are obtained, and this leads to three distinct types of combiners, namely: scanning and selection combiners, equal-gain combiners and maximal-ratio combiners as shown in Fig. 6.3. In the scanning and selection combiners only one a_k is equal to unity at any time, while the others are all zero. The method of choosing which a_k is set to unity distinguishes the scanning from the selection combiner. In the former, the system scans through the possible input signals until one which is greater than a present threshold is found and the system then uses that signal until it drops below the threshold, when the scanning procedure restarts. In the latter, the branch with the best signal-to-noise ratio (SNR) is always selected.

Equal-gain and maximal-ratio combiners accept contributions from all branches simultaneously. In equal-gain, all a_k's are unity and in maximal-ratio each a_k is proportional to the root-mean-square signal and inversely proportional to the mean-square noise power in the k-th branch.

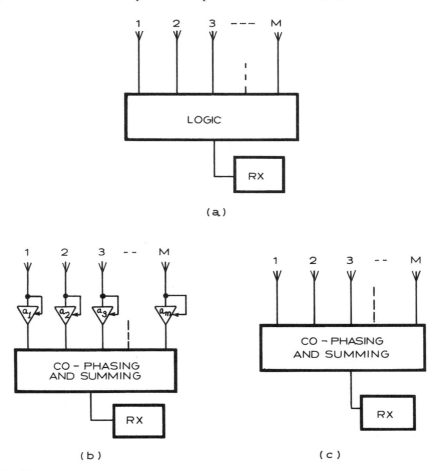

Fig. 6.3. *Diversity reception systems (a) Selection (b) Maximal ratio (c) Equal gain*

While scanning and selection diversity do not use the assumptions (b) and (c), equal-gain and maximal-ratio combining rely on the coherent addition of the signals against the incoherent addition of noise. This means that both equal-gain and maximal-ratio show a better performance than scanning or selection, provided the four assumptions hold. It can also be shown that, in this case, maximal-ratio, among all linear combiners, gives the maximum possible improvement in SNR, the output SNR being equal to the sum of the SNR's from the branches. However, this is not true when either assumptions (b) or (c), or both, do not hold, in which case selection or scanning can out-perform maximal-ratio and equal-gain, especially when the noises in the branches are highly correlated.

6.5 Selection diversity

Conceptually, and sometimes analytically, selection diversity is the simplest of all the diversity systems. In an ideal system of this kind the signal with the highest instantaneous SNR is used and so the output SNR is equal to that of the best incoming signal. In practice the system cannot function on a truly instantaneous basis, and so to be successful it is essential that the internal time-constants of a selection system are substantially shorter than the reciprocal of the signal fading rate. Whether this can be achieved depends on the bandwidth available in the receiving system. Practical systems of this type usually select the branch with the highest signal plus noise, or utilise the scanning technique mentioned in the previous section.

The signal on each branch of a diversity reception system has Rayleigh statistics (eqn. 1) and so the CNR on the k-th branch γ_k has exponential statistics as given by (9). The probability that the CNR is less than any specific value γ_s is given by (10) as

$$\text{prob}\,[\gamma_k < \gamma_s] = 1 - \exp(-\gamma_s/\Gamma)$$

so the probability that the CNR on all M branches is simultaneously less than or equal to γ_s is

$$P_s(\gamma_s) = [1 - \exp(-\gamma_s/\Gamma)]^M \qquad (16)$$

which is the cumulative probability distribution of the best signal taken from the M branches (i.e. selection diversity).

By differentiating, the p.d.f. $p_m(\gamma)$ is given by

$$p_s(\gamma) = \frac{M}{\Gamma}[1 - \exp(\gamma/\Gamma)]^{M-1} \exp(-\gamma/\Gamma) \qquad (17)$$

6.6 Maximal ratio combining

In this method, each branch signal is weighted in proportion to its own signal voltage-to-noise power ratio before summation. When this takes place before demodulation it is necessary to co-phase the signals prior to combining and various techniques for achieving this have been described. Assuming this has been done, the envelope of the combined signal is

$$r = \sum_{k=1}^{M} a_k r_k \qquad (18)$$

where a_k is the appropriate branch weighting. In a similar way we can write the sum of the branch noise power as

$$N_t = N \sum_{k=1}^{M} a_k^2 \qquad (19)$$

so that the resulting SNR is

$$\gamma_R = \frac{r^2}{2N_t}$$

Maximal ratio combining was first proposed by Kahn who showed that if the various values of a_k are chosen as indicated above (i.e. $a_k = r_k/N$) then γ_R will be maximised and will have a value

$$\gamma_R = \frac{(\Sigma r_k^2/N)^2}{2N\Sigma(r_k^2/N)^2} = \sum_{k=1}^{M} \frac{r_k^2}{2N} = \sum_{k=1}^{M} \gamma_k \qquad (20)$$

This shows that the output SNR is equal to the sum of the SNR's of the various branch signals, and this is the best that can be achieved by any linear combiner. The individual γ_k can be written

$$\gamma_k = \frac{r_k^2}{2N} = \frac{1}{2N}(x_k^2 + y_k^2)$$

where x_k and y_k are independent zero-mean Gaussian random variables with equal variance σ^2. It then follows that γ_R is a chi-square distribution of $2M$ Gaussian random variables with variance $\sigma^2/2N = \frac{1}{2}\Gamma$. The probability density function of γ_R is therefore

$$p(\gamma_R) = \frac{\gamma_R^{M-1} e^{-\gamma_R/\Gamma}}{\Gamma^M (M-1)}; \gamma_R \geq 0 \qquad (21)$$

and the cumulative probability distribution function is given by

$$P_M(\gamma_R) = \frac{1}{\Gamma^M(M-1)!} \int_0^{\gamma_R} x^{M-1} e^{-x/\Gamma} \, dx$$

$$= 1 - e^{\gamma_R/\Gamma} \sum_{k=1}^{M} \frac{(\gamma_R/\Gamma)^{k-1}}{(k-1)!} \qquad (22)$$

It is now a simple matter to obtain the mean output SNR from (20) by writing

$$\bar{\gamma}_R = \sum_{k=1}^{M} \bar{\gamma}_k = \sum_{k=1}^{M} \Gamma = M \qquad (23)$$

and thus it can be seen that $\bar{\gamma}_R$ varies linearly with the number of branches M.

6.7 Equal gain combining

Equal gain combining is similar to maximal ratio but there is no attempt at weighting the signals before addition. The envelope of the output signal is given by (15) with all $a_k = 1$

$$r = \sum_{k=1}^{M} r_k$$

and the output SNR is therefore

$$\gamma_E = \frac{r^2}{2NM}$$

Of the diversity systems so far considered, equal gain is analytically the most difficult to handle because the output r is the sum of M Rayleigh-distributed variables. The probability distribution function of γ_E cannot be expressed in terms of tablulated functions for $M > 2$, but values have been obtained by numerical integration techniques using a computer. The distribution curves lie in between the corresponding ones for maximal ratio and selection systems, and in general are only marginally below the maximal ratio curves.

The mean value of output SNR γ_E can be obtained fairly easily as

$$\bar{\gamma}_E = \frac{1}{2NM} \overline{\left(\sum_{k=1}^{M} r_k\right)^2}$$

$$= \frac{1}{2NM} \sum_{j,k=1}^{M} \overline{(r_j r_k)} \quad (24)$$

Now it can be shown that $\overline{r_k^2} = 2\sigma^2$ and $\overline{r_k} = \sigma\sqrt{\pi/2}$. Also, since we have assumed the various branch signals to be uncorrelated, $\overline{r_j r_k} = \overline{r_j}\,\overline{r_k}$ if $j = k$. So (24) becomes

$$\bar{\gamma}_E = \frac{1}{2NM}\left[2M\sigma^2 + M(M-1)\frac{\pi\sigma^2}{2}\right]$$

$$= \left[1 + (M-1)\frac{\pi}{4}\right] \quad (25)$$

6.8 Improvements obtainable from diversity

There are various ways of expressing the different improvements which can be obtained from the use of diversity techniques. Most of the theoretical results have been obtained for the case when all the branches have signals with Rayleigh fading envelopes and equal mean CNR's.

46 Diversity techniques in communications receivers

One useful way of obtaining an overall idea of the relative merits of the various diversity methods is to evaluate the improvement in average output SNR relative to the single branch SNR. For Rayleigh fading conditions this quantity, is easily obtained in terms of the number of branches. However the few dB increase in average SNR which diversity provides is relatively unimportant in most applications. If this was all diversity did, we could achieve the same effect by increasing the transmitter power. Of far greater significance is the ability of diversity to "smooth-out" the fades and reduce the variations in the signal to a level at which the AGC or limiter can cope. In statistical terms we can use diversity to reduce the variance of the signal envelope and this is something which cannot be achieved just by increasing the transmitter power.

To show this effect we need to look at the first-order envelope statistics of the signal, i.e. the way the signal behaves as a function of time. Cumulative probability distributions of the composite signal have been calculated for Rayleigh-distributed individual branches with equal mean CNR's in the previous paragraphs. For two-branch section and maximal-ratio systems the appropriate cumulative distributions can be obtained analytically and for $M = 2$ an expression for equal-gain combining can be written in terms of tabulated functions. The normalised results are of the form

Selection: $\quad P(\gamma < \gamma_s) = (1 - e^{-\gamma_s})^2$ (26)

Maximal-ratio: $\quad P(\gamma < \gamma_s) = 1 - (1 + \gamma_s)e^{-\gamma_s}$ (27)

Equal-gain: $\quad P(\gamma < \gamma_s) = 1 - e^{-2\gamma_s} - \sqrt{\pi \gamma_s}\, e^{-\gamma_s}\, \text{erf}(\sqrt{\gamma_s})$ (28)

and these are plotted in Fig. 6.4.

6.9 The effect of diversity on data systems

In order to assess the effectiveness of a diversity system in respect of data communications we need to determine the reduction in BER that can be achieved from their use. We restrict attention to PSK and FSK systems since these are of major interest.

The form of the BER expressions for PSK and FSK binary systems in the presence of additive Gaussian noise have already been quoted. We have seen how these expressions are modified by fading and we used the technique of writing down $p_e(\gamma)$ and integrating over all possible values of γ weighting the integral by the p.d.f. $p(\gamma)$. We now use the same technique for diversity except that in this case, instead of using $p_\gamma(\gamma)$, we use the p.d.f. of the CNR at the output of the diversity system.

For a selection system the output CNR is given by (17) as

$$p_s(\gamma) = \frac{M}{\Gamma}[1 - \exp(-\gamma/\Gamma)]^{M-1} \exp(-\gamma/\Gamma)$$

So the BER at the system output is the integral of p_e over all values of γ, weighted by this factor. As an example, consider a 2-branch selection system with non-coherent FSK

$$P_{e,2} = \frac{1}{2}\int_0^\infty \exp(-\gamma/2)\frac{2}{\Gamma}[1 - \exp(-\gamma/\Gamma)]\exp(-\gamma/\Gamma)\,d\gamma$$

this is readily evaluated, the answer being

$$P_{e,2} = \frac{4}{(2+\Gamma)(4+\Gamma)} \tag{29}$$

note if $\Gamma \gg 1$ then

$$P_{e,2} \simeq 4P_{e,1}^2 \quad \left(P_{e,1} = \frac{1}{2+\Gamma}\right)$$

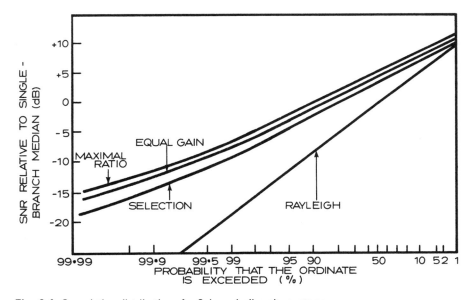

Fig. 6.4. *Cumulative distributions for 2-branch diversity systems*

For a maximal ratio combiner the CNR at the output is given by (21) as

$$P_M(\gamma_R) = \frac{\gamma_R^{M-1}\exp(-\gamma_R\Gamma)}{\Gamma^M(M-1)!}$$

which for a 2-branch system reduces to

$$P_{M,2}(\gamma_R) = \frac{\gamma_R\exp(-\gamma_R/\Gamma)}{\Gamma^2}$$

48 Diversity techniques in communications receivers

and so for non-coherent FSK transmissions

$$P_{e,2} = \frac{1}{2} \int_0^\infty \exp(-\gamma_R/2) \frac{\gamma_R}{\Gamma^2} \exp(-\gamma/\Gamma) \, d\gamma$$

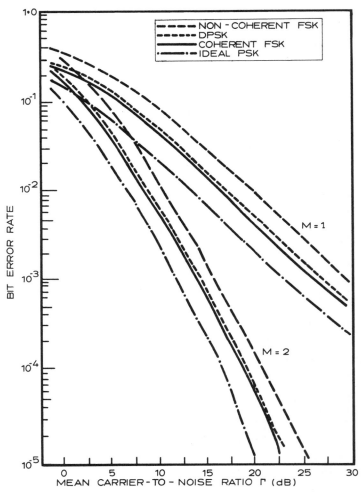

Fig. 6.5. *BER coherent and non-coherent binary systems with 2-branch maximal ratio diversity*

again this is readily integrable, the result being

$$P_{e,2} = \frac{2}{(2 + \Gamma)^2} \tag{30}$$

$$= 2P_{e,1}^2 \quad \left(P_{e,1} = \frac{1}{2 + \Gamma}\right)$$

Turning now to a consideration of coherent detection systems we can restrict attention to combining systems since selection is meaningless in this context and write that in general for these systems the average probability of error is

$$P = \frac{1}{2} \int_0^\infty \text{erfc}\,(\sqrt{\alpha\gamma})\, p_\gamma(\gamma)\, d\gamma$$

There is no convenient closed-form solution of this equation, but using the first term of the expression for $p(\gamma_R)$ from section 6.6 to represent $p(\gamma_R)$ at low error rates we can obtain a solution in the case of maximal ratio when all branches are similar. This solution is

$$P_e = \frac{1}{2(M-1)} \int_0^\infty \text{erfc}\,(\sqrt{\alpha\gamma})\, \frac{\gamma^{M-1}}{\Gamma^M}\, d\gamma$$

$$= \frac{1}{2\sqrt{\pi}} \frac{1}{(\alpha\Gamma)^M} \frac{(M-\frac{1}{2})!}{M!} \tag{31}$$

An exact calculation using the full expression for $p(\gamma_R)$ gives the solid lines of Fig. 6.5 for coherent FSK.

For ideal coherent PSK we would expect to obtain the same curves, but at 3 dB lower SNR since $\alpha = 1$ for PSK compared with $\alpha = \frac{1}{2}$ for FSK. In a similar manner we expect DPSK, for which $\alpha = 1$ to achieve the same error rate at exactly 3 dB less SNR than non-coherent FSK, for which $\alpha = \frac{1}{2}$. To compare coherent and non-coherent systems we can compare DPSK with ideal PSK. It can be shown that for the same error rate PSK requires a smaller CNR on each branch, the relative factor being given by

$$\left[\frac{M!\,(-1/2)!}{(M-1/2)!} \right]^{1/M}$$

Thus at low error rates in Rayleigh fading conditions DPSK is 3 dB worse than PSK, a result which was obtained earlier. However the use of four-branch diversity reduces the difference to about 1 dB, which is the same as the difference under non-fading conditions.

Bibliography

1. BRENNAN, D. G., 1959, "Linear Diversity Combining Techniques, *Proc. IRE,* **47**, 1075–1102.
2. JAKES, W. C. Jnr. (ED): 1974, "Microwave Mobile Communications", John Wiley and Sons, New York.
3. PARSONS, J. D., et. al., 1975, "Diversity Techniques for Mobile Radio Reception", *The Radio and Electronic Engineer,* **45**, 359–367.

Chapter 7
GaAs IC amplifiers for radar and communication receivers

R. S. Pengelly

Allen Clark Research Centre, Plessey Research Limited, Caswell, Towcester, Northants, UK

7.1 Introduction

Monolithic microwave integrated circuits (MMICs) are no longer laboratory curiosities with the introduction of development and pilot production phases within companies world-wide. The benefits to be gained from integrating analogue functions onto single small pieces of GaAs are that sub-systems can be reduced in size by a factor of around ten, parts cost and assembly times are significantly reduced and more complex functions can be realised than those available with hybrid circuits. Much work has also been done in the area of high speed and low power consumption digital GaAs circuits using enchancement mode techniques.

The most popular application for analogue MMICs to date has been in advanced phased array radars where cost, reliability, reproducibility, and performance are all mandatory to the success of such systems. Phased array radars are now being considered for shipborne, airborne and land mobile applications. As far as military systems are concerned phased-array transmit/receive module manufacture, using a number of MMIC chips, is the driving force behind the pilot production facilities that have been set up. Since cost is of prime importance an optimum degree of automated wafer production, chip testing, bonding, module build and testing is required.

MMICs are now being developed for a variety of military, commercial and consumer systems. At relatively low frequencies complete front-ends on single chips are being designed for satellite navigation receivers exploiting the low noise figures that can be obtained from GaAs ICs at frequencies below 2 GHz; at higher frequencies, chip sets for TVRO (T.V. Receive Only) terminals have been designed (Hori *et al.* (1)), circuits for 6/4 GHz satellite transponders (Betaharon and de Santis (2)) are under development whilst applications at millimetre waves have started to generate MMIC development programmes. This chapter highlights one particularly important area of the GaAs IC market – that of low-noise r.f. and i.f. amplifiers for radar and communication receivers. Three specific examples are described giving an indication of the circuit techniques used, the performance achieved and the packaging requirements to meet customer needs.

The three examples are:

(1) A low-frequency (below 2 GHz) low-noise amplifier chip having 20-dB gain with a 2.5-dB noise figure;
(2) A low-noise X-band radar amplifier chip having 18-dB gain and a 3.5-dB noise figure; and
(3) A broadband 0.1-to-6 GHz amplifier chip having 8-dB gain and a 4-dB noise figure. Extensions to the latter design to produce high-gain (i.e. 30-dB or greater) modules containing a small number of chips are also covered.

7.2 Low frequency amplifiers

Most monolithic amplifiers produced to operate at frequencies below 2 GHz or so use one of two design principles – either they employ directly coupled techniques which for all but the simplest amplifiers require level shifting diodes (Hornbuckle and Van Tuyl (3)) or they employ a combination of feedback and lumped element matching (Rigby et al. (4)).

The latter design principle although not offering the highest packing density or lowest power consumption usually results in the best all-round performance in terms of noise figure, power handling and input and output VSWRs. For these reasons it is a technique often chosen for either low-noise front-end preamplifiers for communication receivers or for high i.f. applications.

The IC of Fig. 7.1, for example, is designed to operate between 1200 to 1600 MHz with a gain of 20-dB and a 2.5-dB noise figure for application in a sensitive front-end for reception of global positioning satellite information. The circuit diagram of the preamplifier is shown in Fig. 7.2 indicating the employment of feedback amplification in the first stage using a 1000-micron gate-width, 1-micron gate-length MESFET with the second stage gain being provided by a 500-micron gate width device with the same gate length. By correct choice of feedback resistance, and feedback inductance the terminal VSWRs of the amplifier can be made acceptably low. A feedback amplifier having no other passive matching elements depends for its noise figure on the MESFET and the value of the feedback resistor. In the limit where the feedback resistor is not present the noise figure of the amplifier becomes that of the MESFET working into a characteristic impedance which is usually 50 ohms. A simple means of calculating the noise figure of a MESFET amplifier has been published (Honjo et al. (5)). The feed-back amplifier can be considered as two noise blocks – the noise figure of the feedback resistor is given by:

$$F_1 = 1 + \frac{|1/R_0 + g_m|^2 R_0}{(g_m - 1/R_{FB})^2 R_{FB}} \qquad (7.1)$$

52 GaAs IC amplifiers for radar and communication receivers

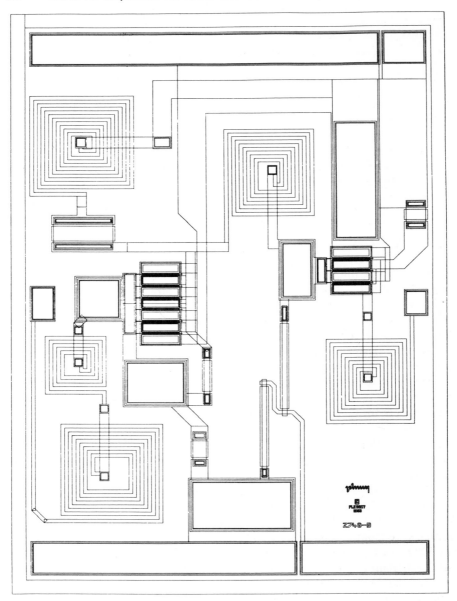

Fig. 7.1 Chip layout of low noise, high gain MMIC amplifier

GaAs IC amplifiers for radar and communication receivers

where R_0 is the drain to source resistance
g_m is the transconductance, and
R_{FB} is the feedback resistor.

If the MESFET has a noise figure of F_2 with a 50 ohm noise figure of F_{50}, then it can be shown that:

$$\frac{F_2 - 1}{F_{50} - 1} \cong 1.5 \tag{7.2}$$

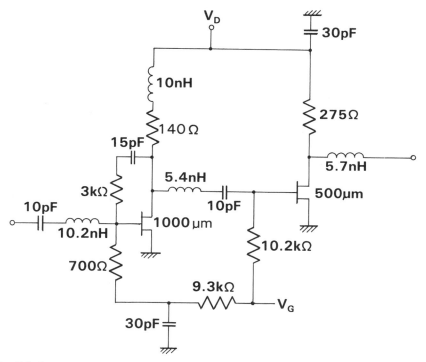

Fig. 7.2 *Circuit diagram of low noise, high gain MMIC amplifier (the chip also contains an input SPDT switch)*

The 50 ohm noise figure is directly related to the noise resistance of the MESFET. From Fukui's theory it can be shown that:

$$R_n \propto 1/g_m^2 \propto \frac{1}{W_g^2}(aL_g/N)^{2/3} \tag{7.3}$$

where W_g is the total gate width, L_g is the gate length, a is the effective channel thickness and N is the carrier concentration in the channel.

The minimum noise figure of the MESFET is given by:

$$F_{min} - 1 \propto fC_{gs}((R_g + R_s)/g_m)^{1/2} \tag{7.4}$$

$$\propto fL_g W_g (g_m(R_g + R_s))^{1/2} \qquad (7.5)$$

Thus by careful choice of the MESFET gate-width and gate-length a low-noise figure with sufficient transconductance to enable effective drain-to-gate feedback to be employed can be produced.

The total noise figure of the feedback amplifier is given by:

$$F_T = F_1 + F_2 - 1 \qquad (7.6)$$

The amplifier of Fig. 7.2 has an input feedback stage followed by a large gate-width FET whereby with simple resistive dumping on its drain a low output VSWR can be obtained. In order to produce a bandpass characteristic, increase the stage gain and improve the input, interstage and output matches some simple passive matching is also used. This consists of square spiral inductors. A single pole, double throw switch is also incorporated on the chip before the amplifier allowing the selection of two different antennae. Chip size is 2.5 by 2.5 by 0.2 mm. The chip is contained in a modified leadless chip carrier.

7.3 Low noise X-band radar amplifier

Recently considerable attention has been given to the subject of designing and realising preamplifiers for phased-array radar receiver applications where noise-figure requirements are near state-of-the-art. An example of such an amplifier is a monolithic circuit designed to operate in the 9.5 GHz radar band shown in Fig. 7.3. Two such chips are designed to operate in a balanced configuration between conventional microstrip Lange quadrature couplers on alumina which then enables a number of chips to be cascaded. In this amplifier design passive components are used to match the MESFETS providing a flat gain over a frequency band ultimately limited by the gain-bandwidth product of the active device. The bandwidth of such an amplifier can be increased by using a large number of matching components but this improvement is clearly gained at the expense of a poorer GaAs IC yield and requires a greater area of GaAs decreasing the number of chips per wafer. Most GaAs ICs are fabricated on substrates which are about 100 to 200 microns thick resulting in component Q's of about 50. Using additional matching components to improve bandwidth will increase the loss of the circuit thus degrading overall gain and noise figure. The most important components for use in such matching networks are capacitors, inductors and transmission lines all of which can be readily integrated into a monolithic circuit. There are two capacitor structures currently in use – these are interdigitated and overlay capacitors. Interdigitated capacitors, utilising the fringing capacitance between parallel strips of metallisation are easier to fabricate accurately but require careful modelling to take into account the parasitics associated with the structure. Overlay capacitors require extra processing steps to deposit the layer of dielectric but have the advantage of fewer parasitics and greater capacitance per unit area.

GaAs IC amplifiers for radar and communication receivers 55

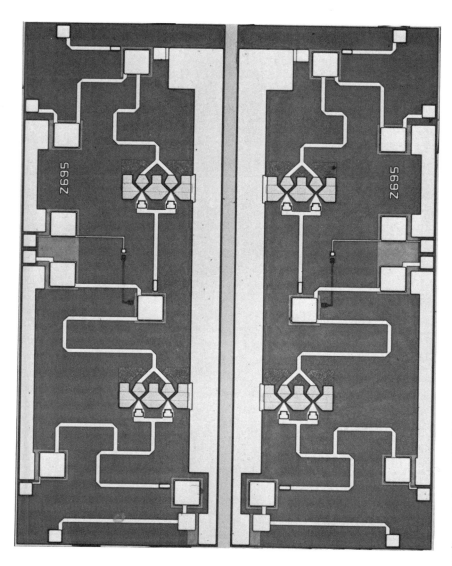

Fig. 7.3 Photomicrograph of monolithic X-band MMIC amplifiers

56 GaAs IC amplifiers for radar and communication receivers

The amplifier shown in Fig. 7.3 is a two-stage amplifier covering the frequency range 7 to 11 GHz. High-impedance transmission lines have been used to simulate the effect of matching inductors and overlay capacitors have been used both as matching elements and as d.c. blocks and bypass capacitors. The gain of the amplifier is 16.5 dB $\pm$ 1.5 dB from 6.6 to 10.6 GHz. The performance of a balanced amplifier consisting of two chips between couplers is shown in Fig. 7.4. Over the 9.5 GHz radar band the gain is 15 dB $\pm$ 0.25 dB, noise figure is 3.8 dB with input and output return losses of better than 20 and 15 dB respectively. Thin film resistors using cermet technology are used whilst the overlay capacitors employ either silicon nitride or polyimide for d.c. blocking, bypass or tuning respectively. Chip size is 4 by 2 by 0.2 mm.

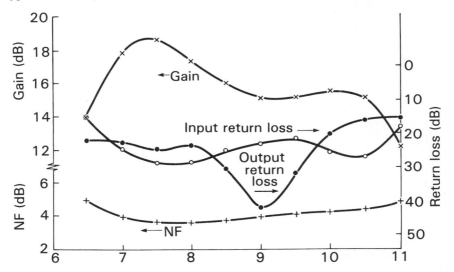

Fig. 7.4 *Performance of balanced X-band MMIC amplifier*

As mentioned earlier, these chips are designed as part of a transmit/receive module for a phased-array radar. By employing such modules the performance ability of future radar systems will be far greater than present day radars. In airborne requirements size and weight are particularly important considerations. Since up to several thousands of modules can be required for any one system the architecture of the module must be chosen to minimise its complexity and thus allow a high yield to be realised in production hence keeping costs low.

7.4 Broadband amplifiers for EW and signal processing

The amplification of microwave signals in a single ultra-wideband amplifier is attractive to various system designers involved in electronic countermeasures

GaAs IC amplifiers for radar and communication receivers

and surveillance as well as those requiring wide-band video and pulse amplifiers. In addition such amplifiers, featuring low noise figures and high 1-dB output compression points, offer large spurious free dynamic ranges which are attractive in a variety of general purpose r.f. and i.f. applications.

The first example of such ultra wideband amplifiers is a 0.1-to-6 GHz monolithic amplifier shown in Fig. 7.5. The MESFET used in this chip is a 900-micron gate-width device where all the sources are connected together using a low inductance/capacitance technique called an airbridge. This is simply a gold plated interconnect line, approximately 5 to 10 microns thick, which is self supporting with an air gap of approximately 10 microns between itself and the IC. The wideband performance depends on the use of drain to gate reactive feedback such that feedback inductance adjusts the S-parameters of the amplifier to produce optimum positive feedback at the upper band edge. Further amplifier performance improvements can be produced by using simple matching networks at the input and output of the MESFET to supplement the feedback network. For monolithic amplifiers these additional networks should be as simple as possible to reduce GaAs acreage as well as being low pass enabling the low frequency performance to be limited only by d.c. blocking and feedback capacitances.

Referring to Fig. 7.5, lengths of high impedance transmission lines have been used to realise inductors. The feedback loop comprises such a length of line in series with a thin film resistor and an overlay nitride capcacitor. The smaller value shunt capacitors used in the input and output matching circuits have been realised using interdigitated structures. The MESFET gate bias is fed in through a high value cermet resistor. Drain bias is fed to the chip via an on chip r.f. choke (the square spiral). The overall chip size is 2.8 by 1.8 by 0.2 mm.

At I_{dss} the typical gain of such a chip is 8 $\pm$ 0.6 dB from 1 to 5.7 GHz. The output VSWR is better than 2:1 over this frequency range whilst the input VSWR is better than 3:1. It is a characteristic of feedback amplifiers that the input is more difficult to match than the output. At low noise bias the measured gain per stage is typically 5.8 $\pm$ 0.6 dB from 0.6 to 6 GHz with a noise figure of around 4 dB over the entire frequency range. Because such an amplifier uses a relatively large gate-width MESFET the 1 dB gain-compression point is high being approximately 20 dBm over the entire band.

The use of feedback also improves the intermodulation distortion – for example, the third-order intermodulation products for this MMIC amplifier at 2 GHz with two signals having equal output powers of 10 mW are 50 dB below the carriers.

The performance of the above amplifier in terms of VSWRs can be improved by decreasing the gate length and overall physical dimensions of the MESFETs thus lowering their parasitic capacitances. Fig. 7.6 shows the response and circuit diagram of a cascade of two-stage amplifier chips employing 900-micron wide, 0.5 micron long-gated FETs each with an overall area which is nearly one-tenth that of the FET used in the previous example. To aid gain flatness

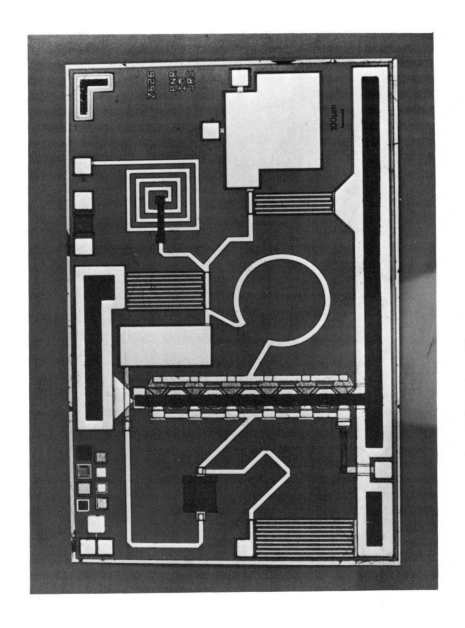

Fig. 7.5 Photomicrograph of monolithic 0.1 to 6 GHz chip amplifier

GaAs IC amplifiers for radar and communication receivers

small amounts of series feedback are included in the FET sources. Each chip contains five silicon nitride capacitors, two 5 pF values in the feeback paths and 5 to 10 pF values for d.c. blocking. Polyimide dielectric overlay capacitors are used for the tuning capacitors where tolerances of ±10% can be achieved. Because of the small overall size of the MESFETs only a single ground is used enabling a simpler feedback topology with no need for through-GaAs vias. The drain bias to each amplification stage is provided by 'on-chip' spiral chokes having resonant frequencies well beyond 10 GHz.

The two stage chip has a bandwidth of 10 GHz with terminal VSWRs which are low enough to allow direct cascading of the chips on a small carrier up to approximately 50 dB of gain. The noise figure of such a multi-stage amplifier is approximately 5 dB over the entire frequency range. Two-stage chip size is 4 by 1.7 by .2 mm.

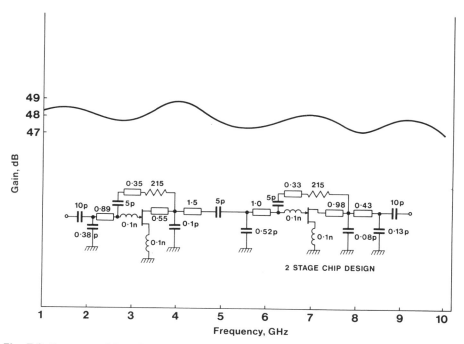

Fig. 7.6 *Response of four, 2 stage chips directly cascaded (2 stage chip design inset)*

7.5 A comparison of circuit techniques

Two basic circuit techniques to realise GaAs MMIC amplifiers have been described in this chapter. Of the methods described the use of frequency controlled feedback most readily provides the widest bandwidth although it can also be used to provide compact narrower bandwidth designs as well (Swift *et al.* (6)). However good quality MESFETs are required to fully utilise this potential. The inherent parasitics of a device limit its performance, in particular the gate-to-drain capacitance of the FET. As mentioned earlier, a device with a high transconductance is also required to provide cascadable terminal VSWRs for a usable level of gain. The necessity of using devices with high transconductance and low inherent parasitics implies a degradation in the overall IC yield, resulting in a higher cost per amplifier. Two other amplifier design techniques which have not been detailed in this chapter should be mentioned. These are:

(1) Actively matched amplifiers (Pengelly, *et al.* (7)); and
(2) Travelling wave amplifiers (Ayasli, *et al.* (8)).

7.5.1 *Actively matched amplifiers*

GaAs MESFETs are designed to give optimum performance (maximum gain and minimum noise figure) when used in common source configuration. The input impedance of a FET in common gate and its output impedance in common drain (source follower) are close to 50 ohms when the transconductance of the FET is 20 mS and when the frequency of operation is well below f_T. Common gate and common drain FETs can thus be used as input and output matching components in MMIC amplifiers. A minimum number of additional passive matching components are then required to produce an amplifier with flat gain over wide bandwidths. If common-gate, common-source and source-follower FETs are cascaded directly with no further matching then a low-pass amplifier with a high level of integration can be achieved (Van Tuyl (9)).

7.5.2 *Travelling wave amplifiers*

There has recently been a resurgence of interest in travelling-wave amplifier design using MESFETs mainly in MMICs. Unlike a conventional circuit the travelling-wave amplifier depends for its operation on artificially producing transmission lines consisting of inductances and the input and output capacitances of the FETs. By adjusting the lengths of these transmission lines between several FETs the phases of the waves on the gate and drain lines can be made such that the waves build up along the structure. Because a transmission line is essentially a broadband component a travelling-wave amplifier can be produced with very wide bandwidths of well over a decade. Because such amplifiers do not provide optimum noise figures they have not been used in conventional radar applications.

7.6 Conclusions

This chapter has given a brief overview of monolithic microwave amplifiers on GaAs with particular emphasis on low-noise preamplifiers for radars and communications and wide-band amplifiers for electronic countermeasures and instrumentation. Two particular design techniques have been examined –

conventional passive matching and feedback – with examples of MMICs produced being given.

7.7 Acknowledgements

The author would like to thank all his colleagues at Plessey Research (Caswell) Ltd. for contributing their work to this chapter, particularly P. N. Rigby, P. D. Cooper, J. R. Suffolk, C. W. Suckling and J. B. Swift. Part of the work described was carried out with the support of Procurement Executive, Ministry of Defence, sponsored by DCVD.

7.8 References

1. HORI, S., KARNEI, K., et al., 1983, *IEEE MTT-S Int. Microwave Symp. Digest*, 59–64.
2. BETAHARON, K. and DE SANTIS, P., 1982, *Microwave Journal*, **25**, 43–68.
3. HORNBUCKLE, D. P. and VAN TUYL, R. L., *IEEE Trans. on Electron Devices*, **28**, 175–182.
4. RIGBY, P. N. et al., 1983, *IEEE MTT-S Int. Microwave Symp. Digest*, 41–45.
5. HONJO, K., SUGIURA, T. and ITOH, H., 1982, *IEEE Trans. on M.T.T.*, **30**, 1027–1033.
6. SWIFT, J. B., SUCKLING, C. et al., 1984, 'An advanced 6/4 satellite downconverter using GaAs integrated circuits', IEE Colloquium on Advances in Microwave Integrated Front Ends, London.
7. PENGELLY, R. S., SUCKLING, C. et al., 1983, 'Monolithic broadband amplifiers for EW applications', IEE Colloquium on GaAs Integrated Circuits, London.
8. AYASLI, Y. et al., 1982, *Elec. Letts.* **18**, 596–8.
9. VAN TUYL, R. L. 1981, *IEEE Trans. on Electron Devices*, **28**, 166–170.

Chapter 8
Integrated optical techniques for acousto-optic receivers

S. M. Al-Shukri, A. Dawar, R. M. De La Rue, G. F. Doughty,
N. Finlayson, and J. Singh*
*Department of Electronics and Electrical Engineering,
The University of Glasgow, Glasgow G12, Scotland, UK*

8.1 Introduction

This chapter is concerned primarily with the use of integrated optical techniques in acousto-optic receivers. More specifically, we consider devices in which a light beam is confined in a waveguide deposited on the surface of a suitable planar substrate (typically a piece of single-crystal lithium niobate) and interacts with a surface acoustic wave (SAW) launched by a specially designed interdigital electrode structure.

In the last ten years, the use of the guided optical-SAW interaction in the integrated optical spectrum analyser (IOSA) has received considerable attention. Work by the authors (De La Rue (1) and (2), Doughty *et al.* (3) and (4)) has addressed some of the key problems and building blocks for IOSA technology and has also attempted to place integrated acousto-optics in relation to other signal processing technologies. The reader is also referred to the recent work (Berg and Lee (5)) for a comprehensive survey of much of the field. The layout of a geodesic lens IOSA is shown schematically in Fig. 8.1.

The inherent reduction by one spatial dimension leads to both advantages and disadvantages for integrated acousto-optics technology in comparison with its bulk counterpart. Some of the arguments are well-known, e.g. the greater interaction efficiency expected for confined optical and acoustic beams and the ruggedness which follows from having more optical and opto-electronic components on the same substrate. But, equally, it has become evident that there are fundamental problems of waveguide quality (and the associated in-plane scattering level), efficient electro-acoustic transduction at frequencies above 1GHz, and economical fabrication of lenses with diffraction-limited performance.

*University of Bath, School of Electrical Engineering, Claverton Down, Bath, UK.

Integrated optical techniques for acousto-optic receivers 63

Furthermore, the extra dimension available with bulk acousto-optics implies much greater flexibility in the realisation of more complex architectures, as demonstrated in another chapter of this book (Bowman *et al.* (6)).

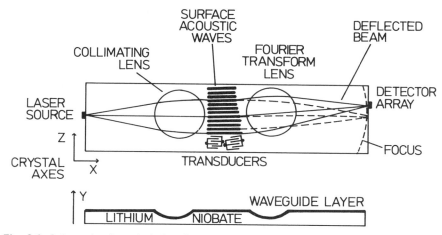

Fig. 8.1 *Schematic of geodesic lens integrated optical spectrum analyser*

8.2 The optical waveguide

It is our belief that the quality of the available optical waveguides has been, and continues to be a crucial stumbling block to the exploitation of integrated optics in acousto-optic receivers. We propose immediately to discard, in a somewhat cavalier manner, the claims of deposited film waveguides (Valette *et al.* (7)). This permits us to concentrate on two rather different diffusion processes in lithium niobate: (i) in-diffusion of thin titanium films (possibly partially or completely oxidised) at temperatures in the range (approximately) 900° C to 1100° C and (ii) the so-called 'proton-exchange' process involving immersion of lithium niobate in benzoic acid (or a range of other 'proton' sources) at temperatures around 200° C. The two processes can also usefully be combined, giving the 'TIPE' process.

Titanium in-diffusion in lithium niobate has been in use for about ten years. In this time, many variations on the basic process have been described and conclusions drawn about the effects of process parameters on waveguide quality (Griffiths and Esdaile (8)). The reader who is not familiar with the literature may well have difficulty in obtaining a clear picture of what is the best procedure to use. We shall refrain here from making categorical assertions, apart from stating that there is still considerable scope for the application of the range of techniques available to the materials scientist to help in assessing the effect of variations in processing (Armenise *et al.* (9)). Attention must be paid both to the base materials, single-crystal lithium niobate and rather impure titanium, and to the solid-state chemical processes which occur during waveguide formation. Correlation of the results of microanalysis with optical measurements of waveguide quality i.e. propagation loss and in-plane scattering (Armenise *et al.* (10)) and Singh (11)) – should eventually lead to waveguides which permit useful dynamic range to be obtained. However, it is not unlikely that detailed process know-how will be of sufficient commercial value that it will not be made available in the open literature.

The proton-exchange process (Canali *et al.* (12) and Al-Shukri *et al.* (13)) has been in use for much less time. The much larger Δn produced by this process ($\Delta n \simeq 0.126$ at $\lambda = 0.6328\,\mu m$ and $\Delta n \simeq 0.9$ at $\lambda = 1.15\,\mu m$) is important for applications in integrated acousto-optics for several reasons. Larger Δn provides stronger light confinement which makes possible better overlap between acoustic and optical fields and also should reduce leakage when the light traverses geodesic lenses with their large deviations from planar geometry. As already demonstrated in a number of publications (Warren *et al.* (14) and Yu (15)) large Δn is also the key to relatively strong planar lens structures which avoid, for instance, use of ion-beam etching of the waveguide surface.

Despite these attractions, the proton-exchange process also has drawbacks (Wong *et al.* (16)) which include large ionic conductivity (Becker (17)), a significant, time-dependent, reduction in refractive index (Yi Yan (18)), a reduced effective electro-optic effect (Becker (19)) and, in our experience, relatively high optical propagation loss and in-plane scattering levels. The refractive index distribution of proton-exchange waveguides can be modified very substantially by relatively moderate annealing processes e.g. half an hour at 325° C. Major modifications to the process are also obtained by using benzoic acid melts containing a small percentage of lithium benzoate and by diffusing titanium into the lithium niobate before carrying out the proton-exchange. Proton-exchange is already providing interesting problems for materials scientists and the use of microanalytical techniques (12, 13). Results obtained on annealed guides using Rutherford back-scattering (RBS) are shown in Fig. 8.2.

8.3 The SAW transducer

The SAW transducers required for integrated acousto-optic receivers must satisfy demanding requirements which are special to this application. Octave bandwidths greater than 500 MHz are desirable and the acoustic beam must be manoeuvered to maintain the acousto-optic Bragg angle over such bandwidths (2). With the use of centre frequencies well above 1 GHz, electron-beam lithographic fabrication becomes desirable or essential, but resistive losses in the relatively thin metallisation ($\sim 1000\,A$) become a major problem and careful design is required in the electrode feed. Recent papers by workers at Plessey (Caswell) report 2.6% per Watt (electrical) diffraction efficiency in a 1.7 GHz centre-frequency structure and 5.5% per Watt in a 1.3 GHz centre-frequency

Integrated optical techniques for acousto-optic receivers

(Kirkby et al. (20), Brambley et al. (21) and (22)). These results are very encouraging and suggest that, with further attention to the key problem of matching to the IDT structure, diffraction efficiencies beyond the reach of bulk acousto-optics may be obtainable.

While, obviously, high diffraction efficiency is desirable in saving optical and electrical power, it is desirable in acousto-optic receivers for more compelling reasons. The advent of significant intermodulation levels arising from non-linearity in the electrical drive amplifier will be more pronounced at, say, the 1 Watt power level than at the 50 mW level and undesirable thermo-optically induced beam-wandering effects are also much easier to avoid.

An important feature of one of the transducer structures described in the work of Brambley et al. (21), is that two sections with different periodicity are joined together with a specified tilt angle of 1.1° between adjacent sections. To avoid writing tilted fingers as 'staircases', the more elegant solution of having two sets of registration marks, with relative tilt between sets, was used.

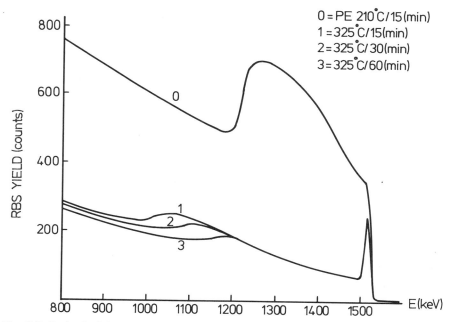

Fig. 8.2 *Rutherford back-scattering measurements on proton-exchanged waveguides on X-cut LiNbO$_3$; without annealing – 0, with various amounts of annealing – 1, 2, 3*

8.4 Input and output coupling of the laser beam

Because of the need for a linear system response, acousto-optic receivers must be operated at low diffraction efficiencies, so that almost all of the light input to the device is wasted. This waste of optical power provides a strong motivation

66 Integrated optical techniques for acousto-optic receivers

for using the interferometric approach (Vander Lugt (23)), since the previously wasted light can be used as the reference beam required to convert the photodetection process from square law to linear dependence on the optical signal field amplitude at the photodetector. In all events it is desirable that light be coupled efficiently into and out of integrated acoust-optic devices. But, for the required spatial frequency resolution, the optical beam must also be several millimetres wide in the region where acousto-optic interaction takes place.

For the IOSA, one approach that has attracted interest is to end-fire light directly from a semiconductor laser into a carefully polished planar waveguide end and then to collimate the beam with a lens placed at its focal distance from the waveguide edge. Geodesic lenses can readily provide the required focal length compatible with the width of the emitting region of typical single-mode stripe geometry lasers.

A quite different approach to the input (and output) coupling problem involves using cylindrical lenses exterior to the acousto-optic device (20, 22). This approach allows the total optical propagation path in the guided-wave device to be reduced to a minimum (in principal no wider than the width of the acoustic beam) with a consequent minimisation of the build-up of in-plane scattered light, as well as of optical propagation loss. The authors are not in a position to judge whether the use of bulk optics represents a major drawback — although it is already reasonably clear that considerable miniaturisation can be achieved so that a suitably folded bulk optics arrangement may occupy a smaller overall length than required in an IOSA substrate. Additional complexity such as is involved in interferometric spectrum analysers and in correlators could possibly favour the integrated optics approach to lenses.

8.5 Lenses

Integrated optical lenses are, arguably, the most intellectually interesting aspect of integrated acousto-optic receivers. Geodesic lenses (Ranganath et al. (24)) have been employed successfully, even though expensive high-precision numerically-controlled machining processes are involved in their manufacture. The Sottini, Righini, Russo theoretical profile (Sottini (25)), in particular, has interesting and somewhat surprising properties (local regions where the radius of curvature tends to zero (1), Doughty (26), Auracher (27)) which provide a challenge to practical implementation. Additionally, it is our view that even with the best profile machining precision available, a numerically controlled polishing process before (and possibly after) waveguide formation is required. In consequence, fabrication costs for geodesic lenses appear likely to be at least in the hundreds of dollars region, even with moderate production volumes (26).

In the last two years, notable progress has been reported on planar lens structures — i.e. ones formed on plane substrates using lithographic definition techniques. Clearly, if adequate optical performance can be obtained, planar

Integrated optical techniques for acousto-optic receivers 67

lenses will be favoured because of the inherently much lower fabrication costs associated with planar technology. The proton-exchange process has been used both in chirp-grating lenses (14) and in a negative-index, concave, step-index change lens (15). It should also be possible to synthesize Luneberg lens profiles using proton-exchange waveguides. Valette (7b) has employed analogue Fresnel lenses in a silicon-based spectrum analyser, achieving diffraction-limited performance. Technically, planar lenses will be favoured once it is clearly demonstrated that the scattered light levels associated with them can be made small enough that overall propagation losses are adequately low over the necessary total optical path length of as much as 10 cm. Some of the basic forms of integrated optical lens are shown schematically in Fig. 8.3.

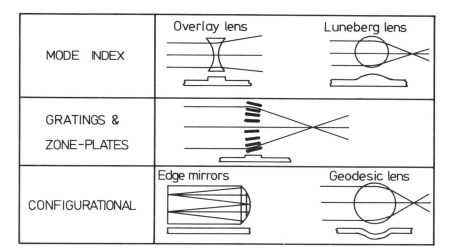

Fig. 8.3 *Schematic of various basic types of integrated optical lens*

8.6 Dynamic range and frequency resolution

Dynamic range and frequency resolution are the fundamental measures of systems performance for acousto-optic recivers of the spectrum analyser type. In a recent paper (Mergerian *et al.* (28)), a dynamic range of 46 dB has been reported in conjunction with 2 μsec integration times, while another paper (Kanazawa *et al.* (29)) reports a 3 dB spot size of 2.7 MHz in conjunction with 400 MHz bandwidth. While these figures for integrated optical devices represent encouraging progress, they certainly fall short of performance available from bulk acousto-optics. Furthermore, assessment of dynamic range and frequency resolution depend both on the performance of the various individual hardware constituents of the spectrum analyser and on how the system is operated – in particular on the nature and number of the rf signals simultaneously present in a device.

Mergerian et al. (28) found that, at $\lambda = 0.6328\,\mu\text{m}$, photo-refractive effects restricted dynamic range to 22 dB, but that increasing the wavelength to the $\lambda = 0.83\,\mu\text{m}$ value available from semiconductor lasers improved dynamic range to 30 dB, it then being limited by the quality of the output spot on a single detector element. While it is tempting to suggest that a further increase in wavelength should further improve performance, there are several competing practical factors which mitigate against this wavelength increase. The further improvement to 46 dB (28) was obtained by re-designing the collimating lens for the semiconductor laser input beam and re-designing the detector array bonding pads.

A feature of both the Mergerian et al. (28) and the Kanazawa et al. (29) devices is that the input collimating lens has a considerably shorter focal length than the output Fourier transform lens – in the latter case the values are 6.6 mm and 52.7 mm respectively. The long output lens focal length thus permitted within the practical constraint of 70 mm substrate length is desirable both because the spot size implied is then of comparable size to the element spacing in the diode array and because de-focusing due to focal surface curvature is minimised. The number of resolved spots obtained (approximately 150) by Kanazawa et al. is still a far cry from the optimistic target of one thousand resolved spots originally set for such devices.

The present authors (10), (11) have investigated two important factors affecting dynamic range – in-plane scattering and acousto-optic third-order intermodulation. In-plane scattering is minimised by an appropriate choice of Ti-diffused waveguide diffusion temperature and time – but further improvements follow from short post-diffusion polishing operations (Vahey (30)). Third-order intermodulation arises from any form of system non-linearity, so that the intermodulation due to the inherent non-linearity of the acousto-optic power relationship (Hecht (31)) represents a minimum estimate of that which occurs in practise – provided that the signal frequencies involved are sufficiently close together. Our results for experiments on an integrated acousto-optic device with two input c.w. signals of equal strength appear broadly to be in agreement with Hecht – but were hampered by the relatively high in-plane scattering background level. There is clearly both scope and the need for measurements on intermodulation levels in broadband beam-steering devices with multiple inputs of different time and frequency characteristics and a range of relative power levels. Our results are shown in Fig. 8.4.

One possible route to greater dynamic range (23) involves greater system complexity – which could justify the use of integrated optical techniques – but only when waveguide and component quality are of a standard not yet achieved.

8.7 Discussion and conclusions, future trends

The material presented above covers the *hardware* aspects of integrated acousto-optics in only a very limited and incomplete manner. *Software*, despite its obvious importance, has not been covered at all! Acousto-optic receivers are attractive because they are 'wide-open' – accepting, essentially continuously, signals over a wide bandwidth (several hundred MegaHertz). By providing an appropriately chosen set of local oscillator frequencies, together with a bank of identical acousto-optic receiver modules, coverage over many GigaHertz of bandwidth is possible. Eventually, all the information from the receiver hardware must be controlled and processed digitally. A major attraction of acousto-optic receivers is that they provide the wideband information in parallel, allowing the flexibility of digital processing to be fully exploited in selecting, *at any time*, which areas of the rf spectrum are to receive particular attention. Given sufficient digital processing power and intelligence in the system, it should be possible substantially to overcome hardware limitations such as intermodulation. Another benefit of the parallel display of the spectrum is that short high-intensity pulses are automatically spread out over a whole range of detector array elements, so that the system is not 'blinded' by such pulses.

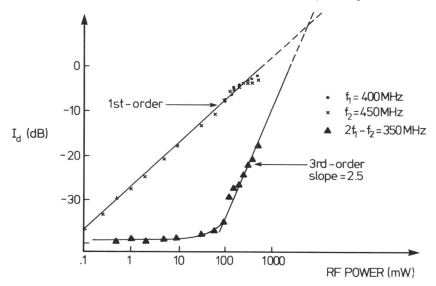

Fig. 8.4 *Experimental observations of two-tone third order intermodulation in a titanium-diffused waveguide acousto-optic device*

It seems worthwhile to pose the question of whether the performance which can reasonably be expected from integrated optical rf spectrum analysers in the near-future has sufficient interest to the systems designer. One can expect 50 dB dynamic range, 200 resolvable spots over 500 MHz bandwidth and moderate power requirements (much less than a Watt). Moderate volume production costs do not appear prohibitive in relation to probable overall system costs. It should also be possible to extend the functions performed by integrated acousto-optics to include convolution and correlation (Tsai (32)). Finally, mention should be made of recent developments in the analogous technology of magneto-optic diffraction by magnetostatic surface waves (Fisher and Lee (33) and Tsai *et al.* (34)). This technology could give wideband signal-processing devices operating directly at frequencies up to 20 GHz with magnetic-field tuning capability over several GigaHertz. However, formidable problems of waveguide quality, integrated lens fabrication and device uniformity need to be solved.

Integrated optic techniques presently lag behind bulk techniques for acousto-optic receivers in several key performance aspects. However, the performance already obtained could prove useful in practical systems and would become very attractive if planar fabrication processes could be implemented for all of the constituent components.

8.8 References

1. DE LA RUE, R. M., 1983, 'Integrated optical spectrum analysers using geodesic lenses on lithium niobate', in (Editors) S. Martellucci and A. N. Chester, 1983, 'Integrated Optics: Physics and Applications', Plenum Press New york, 283–296.
2. DE LA RUE, R. M., 'Integrated acousto-optics: the technology and the competition', in (Editors), Ostrowsky, D. B. and Spitz, E., 1984, 'New directions in guided wave and coherent optics', Martinus Nijhoff, The Hague/Boston/Lancaster, 531–544.
3. DOUGHTY, G. F., DE LA RUE, R. M., SINGH, J., SMITH, J. F., and WRIGHT, S., 1982, 'Fabrication techniques for geodesic lenses in lithium Niobate', *IEEE Trans. CHMT* 5, 205–209.
4. DOUGHTY, G. F., DE LA RUE, R. M., FINLAYSON, N., SINGH, J., and SMITH, J. F., 1982, 'An integrated optical microwave spectrum analyser (IOSA) using geodesic lenses' *Proc. SPIE.* 369, 705–710.
5. BERG, N. J. and LEE, J. N. (Editors), 1983, 'Acousto-optic signal processing: theory and implementation', Mercel Dekker, New York and Basel.
6. BOWMAN, R., GATENBY, P. V., and SADLER, R. J., 1984, 'Acousto-optic correlators', in (Editor), Creasey, D. J., 'Advanced signal processing', IEE Specialist Seminar, Peter Peregrinus, Stevenage, U.K.
7a. VALETTE, S., MORQUE, A., and MOTTIER, P., 1982, 'High performance integrated Fresnel lenses on oxidised silicon substrate', *Electron Lett.* 17, 13.
7b. VALETTE, S., LIZET, J., MOTTIER, P., JADOT, J. P., GIDON, P., RENARD, S., GROUILLET, A. M., FOURNIER, A., and DENIS, H., 1983, 'Spectrum analyser on silicon substrate; a key element for many optical integrated devices', *IEEE Conf. Publ.* 227, 75–76.
8. GRIFFITHS, R. J., and ESDAILE, R. J., 1984, 'Analysis of titanium diffused planar optical waveguides in lithium niobate', *IEEE J-QE*, **QE-20**, 149–159.
9. ARMENISE, M. N., CANALI, C., DE SARIO, M., CARNERA, A., MAZZOLDI, P., and CELOTTI, G., 1984, 'Ti diffusion process in $LiNbO_3$', see reference 2, 623–637.
10. ARMENISE, M. N., CANALI, C., DE SARIO, M., SINGH, J., HUTCHINS, R. H., and DE LA RUE, R. M., 1983, 'Evaluation of planar titanium-diffused $LiNbO_3$ optical waveguides by microanalytical techniques and measurements of in-plane scattering levels', Proceedings of IOOC – 83, Tokyo, Japan, paper 29 A1-3.
11. SINGH, J., 1983, 'Studies concerning the dynamic range of the integrated optic spectrum analyser', Thesis, University of Glasgow.
12. CANALI, C., CARNERA, A., DELLA MEA, G., DE LA RUE, R. M., NUTT, A. C. G., and TOBIN, J. R., 1984, 'Proton-exchanged $LiNbO_3$ waveguides: materials analysis and optical characteristics', *SPIE Proc.*, 460, paper 460-09.
13. AL-SHUKRI, S. M., DAWAR, A., DE LA RUE, R. M., NUTT, A. C. G., TAYLOR, M.,

13. TOBIN, J. R., MAZZI, G., CARNERA, A., and SUMMONTE, C., 1984, 'Analysis of annealed proton-exchanged waveguides on lithium niobate', Topical meeting on integrated and guided-wave optics, Orlando, Florida, U.S.A., PD7-1 to PD7-4.
14. WARREN, C., FOROUHAR, S., CHANG, W. S. C., and YAO, S. K., 1983, 'Double ion exchanged chirp grating lens in lithium niobate waveguides', *Appl. Phys. Lett.*, **43**, 424–426.
15. YU, Z. D., 1983, 'Waveguide optical planar lenses in $LiNbO_3$ – theory and experiments', *Optics Communications*, **47**, 248–250.
16. WONG, K. K., DE LA RUE, R. M., and WRIGHT, S., 1982, 'Electro-optic waveguide frequency translator in $LiNbO_3$ fabricated by proton exchange', *Optics Letters*, **7**, 546–548.
17. BECKER, R. A., private communication.
18. YI YAN, A., 1983, 'Index instabilities in proton-exchanged $LiNbO_3$ waveguides' *Appl. Phys. lett.* **42**, 633–635.
19. BECKER, R. A., 1983, 'Comparison of guided-wave interferometric modulators fabricated on $LiNbO_3$ via Ti indiffusion and proton exchange', *Appl. Phys. Lett.*, **43**, 131–133.
20. KIRKBY, C. J. G., BRAMBLEY, D. R., STEWART, C., and STEWART, W. J., 1983, '500 MHz bandwidth edge-coupled surface wave acousto-optic deflector' paper presented at IEE colloquium on 'Acousto-optic techniques', London, 1st November 1983.
21. BRAMBLEY, D. R., KIRKBY, C. J. G., STEWART, C., and STEWART, W. J., 1983, 'Fabrication of GHz surface-wave acousto-optic deflector (Bragg cell) transducers by electron-beam lithography', paper presented at IEE colloquium on 'Device Physics, Technologies and Modelling for the sub-micron era', London, 23rd November 1983.
22. BRAMBLEY, D. R., KIRKBY, C. J. G., STEWART, C., STEWART, W. J., 1983, 'Edge-coupled, high-efficiency 1.3 GHz SAW acousto-optic deflector', Proceedings of IOOC – 83, Tokyo, Japan, paper 30B3-4.
23. VANDER LUGT, A., 1981, 'Interferometric spectrum analyser', *Applied Optics*, **20**, 2770–2779.
24. RANGANATH, T. R., JOSEPH, T. R., LEE, J. Y., 1981, 'Integrated optic spectrum analyser: a first demonstration', Proceedings of IOOC – 81, San Francisco, U.S.A., paper WH3.
25. SOTTINI, S., 'Geodesic Optics', see reference 2, 545–575.
26. DOUGHTY, G. F., 1984, 'Aspheric geodesic lenses for an integrated optical spectrum analyser', Thesis, University of Glasgow.
27. AURACHER, F., 1983, 'Simple design and error analysis for geodesic lenses', *Wave Electronics*, **4**, 229–236.
28. MERGERIAN, D., MALARKEY, E. C., AND PAUTENIUS, R. P., 1983, 'High dynamic range integrated optical rf spectrum analyser', Proceedings of IOOC – 83, Tokyo, Japan, paper 30B 3-6.
29. KANAZAWA, M., ATSUMI, T., TAKAMI, M., and ITO, T., 1983, 'High resolution integrated optic spectrum analyser', Proceedings of IOOC – 83, paper 30B 3-5.
30. VAHEY, D. W., 1979, 'In-plane scattering in $LiNbO_3$ waveguides', *Proc. SPIE*, **176**, 62–69.
31. HECHT, D. L., 1977, 'Multi frequency acousto-optic diffraction' *IEEE Transactions* **SU-24**, 7–18.
32. TSAI, C. S., 1981, 'Recent progress on guided-wave acousto-optic devices and application', IEE conference publication, **201**, 87–88.
33. FISHER, A. D., and LEE, J. N., 1984, 'Integrated optical signal processing devices employing magnetostatic waves', Topical meeting on integrated and guided-wave optics, Orlando, Florida, U.S.A., TUB4-1 to TUB4-4.
34. TSAI, C. S., YOUNG, D., ADKINS, L., LEE, C. C., and GLASS, H., 1984, 'Planar guided-wave magneto-optic diffraction by magnetostatic surface waves in YIG/GGG waveguides', Topical meeting on integrated and guided-wave optics, Orlando, Florida, U.S.A., TUB3-1 to TUB 3-4.

Chapter 9
Logarithmic receivers
M. Birch
*Allen Clark Research Centre, Plessey Research (Caswell) Ltd,
Caswell, Towcester, Northants, UK*

9.1. Introduction

Several different types of logarithmic amplifiers are used in receivers. The logarithmic circuits under consideration here are high-frequency types intended for operation at video frequencies up to 50 MHz and at receiver intermediate frequencies (i.f) up to 1 GHz. At these frequencies the logarithmic characteristic has to be approximated using multi-stage circuits. Three well-established approximation techniques are examined. Unfortunately logarithmic amplifiers of this type are not easily examined analytically because of the difficulties in dealing with both the non-linear aspects and the approximation techniques involved. For this reason the techniques are discussed in general, simple terms.

Complete logarithmic amplifiers are commercially available. They are commonly based on discrete-component hybrid construction and achieve high levels of perfomance. As systems become more complex, demands on space and weight increase, the ability to tailor construction of receivers to specific equipment requirements becomes more important. However designing and building discrete-component logarithmic amplifiers is a complex and difficult task. Specialised integrated circuits (i.c.'s) are available and these simplify the task of design and construction of amplifiers to meet individual requirements. The use of i.c.'s provide advantages of reproducability, small size, greatly reduced component count and enhanced reliability. The general types of logarithmic i.c.'s both current and future, are indicated.

9.2 Types of logarithmic amplifier

The logarithmic amplifiers commonly used in receivers fall into three broad categories: true logarithmic, successive detection and logarithmic video amplifiers.

A true logarithmic amplifier (1) is normally used in i.f. amplifiers and the

Logarithmic receivers 73

logging process operates directly on the i.f. signal, producing an output at the i.f. amplifier frequency. An important feature of the true logarithmic amplifier is its ablility to maintain near constant group delay over a wide input dynamic range.

The successive detection logarithmic amplifier is also designed to operate at i.f., but provides logged output of the i.f. signal envelope waveform, i.e. the video waveform. Clearly video detectors are an integral function in the realisation of a successive detection amplifiers.

As the name implies the logarithmic video amplifier is designed for purely video-signal operation. A common application is the logarithmic crystal video receiver in which the output signal of a microwave detector diode is directly processed by a logarithmic video amplifier. Because only video is processed logarithmic video amplifiers have much lower bandwidths than the other two types.

Both the true logarithmic and the logarithmic video amplifiers perform the same function; directly acting on the input waveform to provide a logged output of the same signal. But, to meet the different application requirements, their implementations at the circuit level are very different. The true logarithmic amplifier also has some features in common with the successive detection amplifier. By adding a detector to the output of a true logarithmic amplifier, it will do the same job as the successive detection amplifier. However, this would be an inefficient implementation, except where the features of the true logarithmic response were also required. The inefficiency arises from the much higher power dissipation of true logarithmic amplifier which has to process much larger amplitude signals than the successive detection amplifier.

9.3 Achieving a logarithmic response

9.3.1 *General*

Logarithmic circuit receivers are required to operate over large dynamic ranges up to 100 dB, to cover i.f. or video frequencies, whilst maintaining logarithmic law conformance to better than ± 1 dB. Direct methods of obtaining logarithmic characteristics, such as the use of diode feed-back circuits, are not capable of reliably satisfying these requirements. Instead loarithmic-law approximation techniques are commonly employed. These techniques are outlined below, a more detailed account is given by Hughes (2).

The broken line in Figure 9.1. represents the desired logarithmic response,

$$V = A + B \log (e) \tag{9.1}$$

Where e and V are the input and output variables respectively, A and B are constants. The solid lines in Figure 9.1 show a piece-wise linear approximation to the logarithmic curve. For the discontinuities to lie on the logarithmic curve, as shown, the following relationships should be observed:

$$\frac{e_1}{e_0} = \frac{e_n}{e_{n-1}} = K_1 \tag{9.2}$$

and

$$(V_1 - V_0) = (V_n - V_{n-1}) = K_2 \tag{9.3}$$

where e_n and V_n are the input and output variables respectively, at the n-th discontinuity. K_1 and K_2 are constants. From (9.2)

$$e_n = e_0 K_1^n, \tag{9.4}$$

and from (9.3)

$$V_n = V_0 + nK_2. \tag{9.5}$$

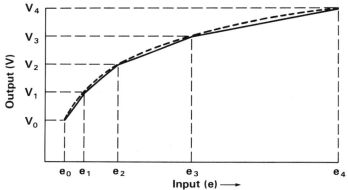

Fig. 9.1. *Piecewise linear approximation to desired response*

Taking logs to base K_1 of (9.4)

$$\log_{K1}\left[\frac{e_n}{e_0}\right] = n \tag{9.6}$$

Substituting for n from (9.5)

$$\log_{K1}\left[\frac{e_n}{e_0}\right] = \frac{(V_n - V_0)}{K_2} \tag{9.7}$$

Rearranging this gives the required form:

$$V_N = K_2 K_3 + \log_{K1}(e_n) \tag{9.8}$$

where

$$K_3 = \frac{V_0}{K_2} - \log_{K1} e_0 \tag{9.8}$$

Away from the discontinuities the approximation deviates from the logarithmic curve. The extent of the deviation can be minimised by increasing the number of linear elements use to cover a given input dynamic range.

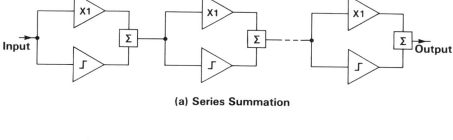

(a) Series Summation

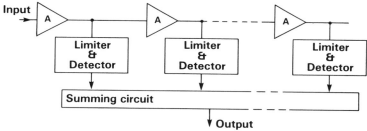

(b) Parallel Summation using Detectors

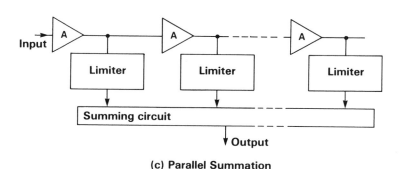

(c) Parallel Summation

Fig. 9.2. *Configurations used in logarithmic amplifier circuits*

Inspection of basic relationships required of the approximation indicates that this response can be realised by cascading linear amplifiers together with a limiting action at each stage. Limiting action is fundamental, it determines the position of the discontinuities on the piece-wise-linear approximation. Three basic configurations are illustrated in Figure 9.2 and are described below. The functional blocks are assumed to have ideal responses.

9.3.2 Series summation

The true logarithmic amplifier is based on this principle, (1) the basic configuration consisting of cascaded stages as in Figure (9.2a). Each stage has a limiting amplifier having a small-signal gain of A in parallel with a unity-gain amplifier; the outputs of the two amplifiers are summed. As each stage limits with increasing input level the gain of the configuration is divided by a factor $(A + 1)$, and the change in output level between each stage limiting is function of the limiting level V_L. Thus K_1 equates to $(A + 1)$ and K_2 to $V_L - V_L/(A + 1)$. Each stage represents a section of the approximation.

9.3.3 Parallel summation using detectors

Figure 9.2b shows a configuration that behaves in a similar manner, but because the summation is performed by taking the outputs in parallel no unity-gain bypass stage is required. However if the individual paths to the summation circuit are linear (apart from the limiting action), the resulting characteristic will deviate from that described above. This arises because additional terms appear in equation (9.4).

$$e_n = e_0(K_1^n + K_1^{n-1} + K_1^{n-2} + \ldots) \tag{9.9}$$

Thus adding to the error in logarithmic law conformance.

The configuration shown in Figure 9.2b is that of the successive detection amplifier and here the limiting element is shown incorporated with the video detector. In commercial amplifiers the detector is made non-linear and when optimised this characterisitic compensates for error caused by parallel summation.

9.3.4 Parallel summation

Another arrangement similar to that above is illustrated in Figure 9.2c, which is the configuration on which logarithmic video amplifiers are based. Here the detector element is replaced by a non-linear limiting amplifier, again non-linear to improve the overall logarithmic law where parallel summation is used.

9.3.5 Computer simulation

The introduction of non-linear elements (in addition to limiting) for the successive detection and logarithmic video amplifier configuration renders them very complex to evaluate analytically. Alternative techniques such as numerical methods or computer aided design are used. Particularly useful are non-linear circuit analysis programs.

For example the non-linear transient analysis of SPICE (3) can be used to simulate these logging configurations. This can be used at two levels. Simple circuits can be used to form macromodels of functional blocks, enabling basic operation of a configuration to be evaluated. At a more complex level, during design, complete circuits including parasitic components can be simulated.

9.4 Circuit considerations

So far the linear and non-linear elements required to produce logarithmic responses have been assumed to be ideal. There are many aspects of circuit perfomance in practical realisations which affect the operation of logarithmic amplifiers. The most important of these are outlined below.

9.4.1 *Amplifier gains*
The main factor affecting the choice of gain is the desired logarithmic-law accuracy and complexity constraint (2). A greater number of stages used for a given dynamic range will yield improved logarithmic accuracy, but this will be at the cost of circuit complexity. The gain range per stage used in most commercial log-amplifiers is 10 dB to 15 dB.

Absolute gain of the unity gain amplifier of the true-log circuit is particularly important, errors here mean that equation (9.2) is no longer valid and furthermore, cascading effects result in error accumulation. To obtain log-law accuracies to better than ± 1 dB, in a 60 dB, six stage amplifier, means keeping gain to 0 dB $\pm$ 0.25 dB at all operating frequencies, and signal levels.

Gain matching is required for all types, again if this is not so, equation (9.2) no longer applies. Also it is necessary where non-linear elements are used in the successive detection and logarithmic video types, that the linear gains are matched to the non-linear characteristic elements.

9.4.2 *Limiting*
It is the task of the limiting sections to ensure that once a stage has completed its contribution to the output signal, that further increase in input signal level does not affect the output level from that stage. A limiter that exhibits differential gain after reaching its limiting level will introduce errors. Again the most demanding application is series summation of the true-logarithmic circuit where limiting stages have to tolerate high levels of overdrive. Practical limiters also deviate from the ideal at the transition from amplifying to limiting, the transition is smooth rather than discontinuous. In general this feature helps to produce a more accurate overall logarithmic response.

To avoid long recovery-time problems after overdrive, transistor saturation should be prevented by careful circuit design.

9.4.3 *Propagation delays*
In parallel summation configurations the signal delays on each sum line are different. The differential delays result from the propagation delay through the cascaded gain stages. Provided that the gain stages exhibit propagation times that are much less than the video pulse rise times required this effect will not give rise to problems. Where this is not the case, video pulse rise times will be degraded, it is possible to recover the situations by adding compensating delays in the summation lines.

78 Logarithmic receivers

In the series-summation case absolute propagation delay is not a problem, however it is important that delays of the unity gain and limiting amplifiers are matched.

9.4.4 *Bandwidth of amplifying stages*
Due to the cascading, the bandwidth of the total circuit will be lower than that of the individual stage. The amount of bandwidth loss depends on the number of stages used and the particular type of amplitude response of the individual amplifier. To ensure a flat amplitude response over the i.f. band of interest, these losses must be taken into account in design for i.f. operation. Two more factors, large-signal bandwidth and phase response, are important for the true logarithmic amplifier. Ideally group delay will be constant over the frequency range and the full dynamic range. To approximate to this, wider bandwidths for a given i.f. are required in addition good r.f. limiting performance and close group-delay matching between the limiting and unit gain elements. The large-signal bandwidth is necessary in the unity-gain elements to accommodate the serially summed i.f. signal. The more demanding bandwidth requirements of this configuration are reflected in lower frequency of operation and higher power dissipation when compared with a successive detection type designed on similar technologies.

9.4.5 *Noise considerations*
The self-noise of the i.f. logarithmic circuits should be well below the start of logging range. As a practical design rule, the mean self-noise power should be at least 20 dB below the level which would cause the last stage to limit, i.e:

$$[(KTB_n)' + F + 20\,\text{dB}] < [P_L - NG] \tag{9.10}$$

$(KTB_n)'$ is fundamental noise power in dBm. Where K = Boltzman's constant, T = temperature in °K, B_n = equivalent noise bandwidth in Hz, P_L = limited output in dBm, N = number of stages, G = stage gain in dB and F = noise figure in dB. So that

$$F < [P_L - (NG + (KTB_n)' + 20)]\,\text{d}B \tag{9.11}$$

The logarithmic video amplifier is usually specified in terms of tangential sensitivity, which is a qualative criteria which can be equated to amplifier noise figure (4).

9.5 Integrated circuits

9.5.1 *Successive detection log amplifier*
Broadband gain, limiting and detection are functions readily producible as an integrated circuit. The matching of components required for logarithmic cir-

cuits is an inherent feature of i.c.'s, and it minimises the need for trimming to obtain good logging performance. All the major circuitry including bias and r.f. decoupling is integrated, greatly reducing the number of components required to build the complete logarithmic amplifier. Circuit techniques are based on differential-pair configurations. The amplifying stages incorporate series feedback to stabilise gain. An offset differential pair can be used to form a low-level detector circuit, which has a non-linear response that provides an improved logging characteristic. Limiting action is inherent in both circuits. Commercially available i.c.'s work with a nominal stage gain of about 12 dB, both single and dual stage building blocks in TO5 packages are available. These circuits produce a current output which is fed to an external low-impedance summing and video filtering circuit.

The operating bandwidth capability of the circuits is determined by the i.c. technology used for their fabrication. Early log i.c.'s were designed on processes yeilding transistor f_T's of about 800 MHz, and these circuits will operate at i.f.'s up to 100 MHz. Later designs are built on a more advanced bipolar technology, have four-micron geometries providing f_T's of about 2.5 GHz, capable of operating in logarithmic i.f. circuits up to 200 MHz.

9.5.2 Logarithmic video amplifiers
The early i.c.'s available for logarithmic video applications provided some of the elements required. These were the non-linear limiting and summing elements. A complete logarithmic amplifier is built by adding video operational amplifiers, with their gain adjusted to match the non-linear elements. To reduce the number of linear amplifier stages to be added, a modified version of the parallel summation configuration is used. The non-linear element is designed to operate over an input dynamic range of 15 dB. But by feeding two sections in parallel, one directly and the other via a 15 dB attentuator, the combined block acts as a 30 dB stage. Four stages have been integrated providing a potential 120-dB dynamic range.

Later developments have concentrated on raising the levels of integration by incorporating the matched linear amplifiers together with the non-linear stages onto a single chip, so that ultimately the whole log-video circuit is available in one i.c. package.

9.5.3. True logarithmic-amplifiers
Only recently has a true-logarithmic i.c. become commercially available. This circuit, described by Barber and Brown (5) consists of a limiting amplifier together with a large-single handling unity-gain amplifier, providing 10 dB of small signal gain; both are broadband. An advanced four-micron ion implanted i.c. process, with transistor f_T's of 5 GHz is used. The 3 dB small-signal bandwidth is almost 500 MHz and true-logarithmic operation can be achieved at i.f.'s up to 100 MHz, whilst maintaining excellent phase.

9.6 Future integrated circuits

Continuing development of new silicon integrated circuit technologies provides the basis for logarithmic amplifier i.c.'s in near future. Technologies now emerging have minimum transistor-emitter geometries of 3 microns and below, they produce high quality inter-level metal nitride dielectric capacitors, use oxide isolation and employ polysilicon emitters. These technology advances improve transistor high-frequency performance and reduce effects of unwanted parasitic circuit components. Further technology advances are anticipated in silicon and also in gallium arsenide with the development of more mature processes.

Designs currently in development are producing limiting amplifiers with bandwidths in excess of 1 GHz. This level of performance is compatible with the operation of successive detection amplifiers at i.f.'s up to 600 MHz and of true logarithmic amplifiers up to 300 MHz. Future designs are expected to reach i.f. operation at frequencies up to 1 GHz and 500 MHz for successive detection and true logarithmic types respectively. To take advantage of these developments improved integrated circuit packaging will be used. Leadless chip carriers are now available, these offer reduced physical dimensions, enhanced high-frequency performance and can be surface mounted.

9.7 References

1. WRONCOW, A., and CRONEY, J., September 1966, 'A True I.F. Logarithmic Amplifier using Twin-Gain Stages', *The Radio and Electronic Engineer*, 149–155.
2. HUGHES, R. S., 1971, 'Logarithmic Video Amplifiers', Artech House, Inc., Dedham, Massachusetts, U.S.A.
3. NAGEL, L. W., 1975, 'SPICE 2: A Computer Program to Simulate Semiconductor Circuits', Electronic Research Laboratory, University of California, Berkely, Report No. ERL-M520.
4. LUCAS, W. J., 1966, 'Tangential Sensitivity of Detector Video System with r.f. Preamplification', IEE Proc., Vol. 113. 1321–1330.
5. BARBER, W. L. and BROWN, E. R., 1980, 'A True Logarithmic Amplifier for Radar I.F. Applications, IEEE JSSC, Vol. SC-15, 291–295.

Chapter 10
CCD processors for sonar

J. F. Dix and J. W. Widdowson
Admiralty Research Establishment, Portland, UK

10.1 Introduction

Sonar systems may be divided into two broad categories identified primarily by the size of the transducer array. Large-array sonars, operating at low frequencies, provide relatively large surveillance ranges out to many nautical miles, whereas the small-array sonar, operating at higher frequencies to achieve comparable, or in some cases better, angular resolution, can only provide ranges out to, at most, about one nautical mile. The size of the array, and consequent overall platform dimensions, can then have a marked influence on the technology employed in implementing the system (Dix (1)).

Inherent in the active role of a small-array sonar is a need for high-range resolution either to achieve a specified degree of echo detail or to meet performance requirements under reverberation-limited conditions. Such a high-range resolution entails the use of wide-bandwidth signals (Arthur et al. (2)).

Depending on the precise nature of the use of the sonar there is often a considerable operational advantage, especially when space is limited, in having a facility to modify critical performance parameters in response to the particular environmental problems encountered (2). For example rough seas create a need for stabilised spatial processing. Such flexibility is feasible when system processing is based on the manipulation of sampled versions of the signals involved because then, the processing can be readily interfaced with a digital computer to provide a programmable system of spatial and signal processing. The need for stabilisation, together with the requirement to handle wide-bandwidth signals, leads to an argument in favour of working in the time domain ((2) and Curtis and Wickenden (3)) to implement programmable spatial processing. Equally, the requirement for wide-bandwidth signals leads to the consideration of a time domain approach to programmable signal processing.

10.2 Time domain programmable spatial and signal processing

10.2.1 *Spatial processing*

Spatial resolution can be obtained from an array of transducer elements by suitably amplitude weighting and time delaying the element outputs to form steered beams. Referring to the network illustrated in Fig. 10.1, if a linear array of N transducer elements is connected to the N port interface, then the output $y(t)$ at port Z, for an incoming plane wave $x(t)$, can be represented as

$$y(t) = \sum_{n=0}^{N-1} a_n x(t - n\tau') \qquad (10.1)$$

where

$$\tau' = \tau - \frac{d \sin \theta}{c},$$

d is the element spacing, c is the velocity of propagation, θ is the angle of the plane wave to the normal to the array, and a_n is the amplitude weighting coefficient for the nth element.

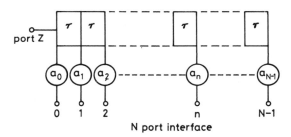

Fig. 10.1 *Delay and amplitude weighting coefficient network*

Alternatively, the output at port Z can be expressed in terms of the angle θ, as an array response or directivity pattern,

$$A(\theta) = \sum_{n=0}^{N-1} a_n \exp(-j 2\pi f n \tau') \qquad (10.2)$$

where f is the frequency of the incoming wave $x(t)$.

Clearly, $A(\theta)$ is a maximum when $\tau' = 0$ and $\tau = d \sin \theta / c$, that is when the time taken for the wave-front to travel between the elements is equal to the network incremental delay time τ thus producing a beam steered at angle θ to the normal to the array.

It is to be noted at this point that equation (10.1) represents discrete convolution and equation (10.2) represents a discrete Fourier transform for a processing scheme essentially sampling the signal waveform in space.

CCD processors for sonar 83

10.2.2 Signal processing

The basis of signal processing is the realisation of a suitable filter response to enhance the presence of a wanted signal waveform in a background of noise and/or interference. Again, referring to the network of Fig. 10.1, if a signal $x(t)$ is fed into port Z and the output $y(t)$ is derived from the summation of the N ports at the N-port interface then,

$$y(t) = \sum_{n=0}^{N-1} a_n \, x(t - n\tau) \tag{10.3}$$

Alternatively, the output can be expressed in terms of frequency as,

$$y(f) = \sum_{n=0}^{N-1} a_n \exp(-j2\pi f n\tau) \tag{10.4}$$

As in the case of equations (10.1) and (10.2) for spatial processing, equations (10.3) and (10.4) for signal processing have the same characteristics, representing discrete convolution and a discrete Fourier transform respectively. In fact, the network of Fig. 10.1 with the N ports summed corresponds to the familiar configuration of a transversal filter, whereby any impulse response, and hence frequency response, can be achieved by suitable selection of delay time τ and amplitude weighting coefficients a_n. The signal processing task is thus a case of designing the filter to maximise the received signal to background noise ratio from a knowledge of the particular signal waveform characteristics involved. For example, if the impulse response $a_n = x(-t)$ for a signal waveform characteristic $x(t)$, then the transversal filter becomes a matched filter chosen to maximise the signal to background-noise ratio for noise with a Gaussian distribution.

For the particular case of a linear-frequency modulated sweep (chirp) impulse response, the matched filter can be made to perform as either a pulse-compression receiver directly (Dicke (4)) or a spectrum analyser via the chirp-Z transform (Rabiner et al. (5), Buss et al. (6) and Widdowson (7)).

10.2.3 Programmable processing

It has been seen that a network of delay elements and amplitude weighting coefficients serves as a fundamental constituent in both spatial and signal processors operating in what is essentially the sampled-time domain, although of course equivalent operations in the discrete-frequency domain can readily be derived as shown in equations (10.2) and (10.4). Now, if the incremental delay time and the amplitude weighting coefficients can be arranged to be under direct control of a computer, then the network of Fig. 10.1 becomes the basis of time domain programmable spatial and signal processing satisfying the requirements outlined at the end of Section 10.1.

At this point, it is also worthwhile to note that the transversal filter configuration of the network in Fig. 10.1, when under computer control, lends itself very readily to the concept of adaptive processing (Widrow et al. (8)).

10.3 Implementation of programmable spatial and signal processing

10.3.1 Choice of technology

Having specified the basic processing network, the next stage is to discuss the more practical aspects of implementation, bearing in mind such constraints as physical size and operating temperature in the case of a compact sonar, where packing density of components is not only determined by their dimensions but also by the power density and resulting temperature rise within the assembly. In addition, to cope with reverberation-limited conditions, there is a need for essentially linear operation (2) over a reasonably wide dynamic range of the order of 60 dB and this can have a considerable bearing on the mode of operation and the type of device employed.

Considering, for example, the spatial processing role of the network of Fig. 10.1, if for instance digital processing were to be entertained, then the process of analogue-to-digital conversion has to be applied to each and every transducer element with a 12-bit resolution readily achieving better than 60 dB dynamic range. Although perfectly feasible, the compact nature of the requirement poses severe problems as far as operating speed and power dissipation are concerned especially when the relatively high operating frequency of the small-array sonar is considered. It was such problems as these that prompted consideration of the charge coupled device (c.c.d) for such a role. The c.c.d. (Boyle and Smith (9)) operates on analogue signal samples directly and, because it is based on m.o.s. technology, can be designed for compact low-power operation. However, being analogue it does, of course, demand more specialist design expertise and circuitry to achieve the necessary stability of performance and dynamic range, than the digital approach.

10.3.2 Sequence of processing operations

The basic operations in a sonar system are spatial processing or beamforming, signal processing, and data processing. Data processing, the last in the sequence, is there to present the sonar-derived information in the best format for ready assimilation by the operator. Among several options the sonar designer has a choice in which order to place beamforming and signal processing and whether to take signal samples from the actual operating frequency range or within a frequency range close to baseband. The choice finally depends on signal-to-noise considerations and processing device-performance limitations. System complexity is usually eased if signal sampling is performed within the normal operating frequency range, but the number of samples involved for some forms of processing can then be too large for devices currently envisaged. To overcome this problem sampling can be performed at baseband resulting in a reduction in sampling rate and the overall number of samples required to represent the signal. However, frequency conversion to baseband can involve a noise penalty together with the generation of undesirable inter-modulation products and it is therefore essential to ensure that the signal before frequency conversion is band-limited and noise free. In addition, any other implications of frequency conversion on overall system performance have to be very carefully assessed (see Section 10.5.1).

The spatial processing or beamforming proposed involves the coherent addition of transducer-element outputs resulting in an array gain and hence an enhanced signal-to-noise ratio at the output of a beam-former. Also the beamformer version of the network in Fig. 10.1 is simple enough to enable signal sampling to be performed within the normal operating frequency range and still represent a sufficiently compact architecture to process all the signals associated with a number of transducer elements. The signal processing version of Fig. 10.1 on the other hand, tends to be more complex in order to get the performance required and sampling has to take place at baseband to keep device geometries within reasonable limits.

These considerations lead to the sequence: spatial processing or beam-forming (at operating frequency), frequency conversion, signal processing (at baseband), and data processing, for the compact sonar system.

10.4 Programmable beam-forming and steering devices

In the beam-forming and steering role of the network in Fig. 10.1, it is convenient to implement the amplitude-weighting coefficient function a_n using a programmable-gain amplifier with sufficiently wide dynamic range to cover the dual function of array element-amplitude shading and signal-amplitude con-

Table 10.1 *Performance summary of beam-forming CCD's designed by the Wolfson Microelectronics Institute, Edinburgh*

Current device parameters	Transmit (WM2020-TDL)	Receive (WM2210-TDI)
Operating temperature	$-20°$ C to $+70°$ C	
Maximum clock frequency	2 MHz	10 MHz
Through gain	-5 dB	$+3$ to $+4$ dB
Dynamic range	106 dB/$\sqrt{\text{Hz}}$	
Device matching	better than 0.5 dB	
Tap matching	better than 0.2 dB	
Total harmonic distortion	-40 dB	
Power dissipation	400 mW	250 mW
New device	WM2030	WM2220
Architecture	Equivalent to 9 × WM2020	Equivalent to 9 × WM2210
Parameters as above except for power dissipation	600 mW	500 mW

86 CCD processors for sonar

ditioning. The signal dynamic range can then be controlled to fall within the range handled by the beam-steering devices. These devices then reduce to comparatively straightforward charge-coupled delay line structures where the facility of programmable beam-steering is achieved by using computer control of a clock frequency to govern the incremental delay time τ. Reasonably extensive descriptions of these devices and their performance can be found in Dix *et al.* (10).

Because of the simplicity of the c.c.d. beam-steering structure, recent developments have been aimed at increasing the device density and facilities on chip. Thus improved devices incorporate techniques for switching delay lines to allow steering either side of the normal to the array plus extra circuitry to include clock generation on chip (Thomson (11) and Thomson and Donaldson (12)). Additional improvements (Denyer *et al.* (13) and Dean *et al.* (14)) have involved even further increases in packing density of these circuits to the point where a potential package-count reduction of better than 4:1 is possible compared with the early beam-forming architectures reported in (10). A summary of performance parameters for existing and future designs of c.c.d. beam steering packages is given in Table 10.1.

10.5 Programmable signal processing devices

Sonar signal processing requirements usually cover both active and passive modes of operation and to meet these requirements in the case of the small-array

Table 10.2 *Performance summary of CCD/MOS monolithic transversal filters designed by the Wolfson Microelectronics Institute, Edinburgh*

Parameter	Early version (WM2111)	Latest version (WM2130)
Operating temperature	$-20°$C to $+70°$C	
Filter points per chip	256	
Through gain	within 0.2 dB	
Weighting accuracy	within 0.2 dB	
Charge transfer inefficiency/point	10^{-3}	$0.4.10^{-3}$
Dynamic range/filter point	90 dB/$\sqrt{Hz}$	108 dB/$\sqrt{Hz}$
Total harmonic distortion	-40 dB	-45 dB
Maximum clock frequency	2 MHz	200 kHz (note)
Power dissipation	280 mW	250 mW

Note: The reduction in maximum clock frequency in the latest version (WM2130) is caused by the introduction of a time-multiplexed signal zero to increase the device dynamic range.

CCD processors for sonar

sonar it often becomes necessary to resort to matched-filter techniques (2). For active ranging the matched filter operates as a pulse compression receiver whilst for active Doppler processing and passive operation, the matched filter is set up to function as a spectrum analyser. The programmable transversal-filter version of the network in Fig. 10.1 (Section 10.2.2) has also been fabricated as a c.c.d. In this case a 256 point charge-coupled delay-line structure, associated m.o.s.t. multipliers, m.o.s. capacitors to store the values of amplitude weighting coefficients, and a shift register to implement a serial interface with a computer have all been integrated on a single chip. The possibility of cascading up to eight devices caters for programmable matched-filter operation involving waveforms with bandwidth-time products (BT) up to 1024. This device has been described fully in the literature ((10) and Denyer and Mavor (16)) and Table 10.2 summarises its salient design features together with those of a later improved version, WM 2130 in Milne et al. (17).

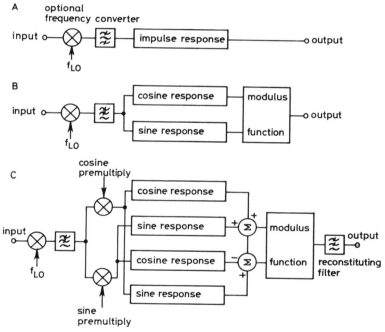

Fig. 10.2 *System architectures*

10.5.1 *Matched-filter system architectures*

When designing c.c.d.-based matched-filter type processing systems careful consideration must be given to the system architecture in order to achieve optimal performance. Essentially there are three system choices (Fig. 10.2), all of which are designed to minimise losses due to phase mismatch between the input signal and the stored impulse–response waveform.

88 CCD processors for sonar

System A functions at either the sonar operating frequency or at some intermediate-frequency greater than the signal bandwidth, with phase losses decreasing the higher the frequency. This system is simple and does not have any requirements for modulus circuitry or channel matching. The system is however inefficient in terms of store size and requires a high operating speed. It is also greatly affected by charge-transfer inefficiency (ε) being unusable for $BT\varepsilon >$ 0.25 (Denyer (18)). Nevertheless this architecture is well worth considering for a matched filter because of its simplicity.

System C represents the optimum choice for c.z.t. based systems as maximum use can be made of the processor by utilising the sliding chirp-z transform to continuously process incoming data. With this architechture both frequency and time domain processing can be accomplished by the addition of an initial frequency converter and the use of suitable pre-multiplier signals. This system does, however, require the addition of a signal reconstituting filter if sampling errors (Clarke (19)) are not to reduce the performance by up to 4 dB. In processors which are highly programmable, it may not be possible to use a reconstituting filter because of the differing signal bandwidths being used. Sampling errors may then preclude the use of architecture C and system B then becomes the obvious choice as sampling error loss is reduced to about 1 dB. System B also shows similar performance in the presence of charge-transfer inefficiency to system C and both are much improved over system A. Thus system B architecture is suited to the active ranging, pulse compression application and system C to the Doppler and passive processing, spectrum-analyser applications.

10.6 Conclusions

A complete breadboard four channel assembly of programmable beam-steering and signal processing has been tested at sea. Preliminary experiments have confirmed successful operation of computer-controlled beam-steering, signal-amplitude conditioning, and active and passive processing under conditions representing a fully integrated system. Having demonstrated the viability of such an assembly, work is now progressing on a second version, adequately engineered, to stand the rigours of a comprehensive sea trials programme and operate reliably for long periods of time.

As far as the future role of the c.c.d. in compact sonars is concerned, this is open to speculation. The c.c.d. development work summarised in this paper is the result of a six year programme of design, evaluation and improvement aimed at the sonar application. Present designs are considered to be close to the limits of what can be achieved in charge-coupled technology, and it remains to be seen whether equivalent digital devices, with improved stability and dynamic range, will eventually appear as a result of future programmes of work in VHSIC/VLSI technology. In the meantime it is felt that the existing c.c.d.-based system referred to above will continue to provide a very useful research tool in investigating the application of programmable spatial and signal processing to improve the performance of the small-array compact sonar.

10.7 Acknowledgements

This work has been carried out within the Procurement Executive, UK Ministry of Defence. The c.c.d.'s and some of the related circuitry referred to in this paper were designed by P. B. Denyer, R. S. Thomson, and J. R. C. Reid of the Wolfson Microelectronics Institute, Edinburgh.

Copyright © Controller HMSO London 1985

10.8 References

1. DIX, J. F., 1978, 'Initial Sonar System Considerations'. Digest of IEE colloquium on Modern Signal Processing Technology for Sonar, London, Digest 1978/52.
2. ARTHUR, J. W., DIX, J. F., HARLAND, E., WIDDOWSON, J. W., DENYER, P. B., MAVOR, J., and MILNE, A. D., 1979, 'Large Time-Bandwidth product CCD Correlators for sonar', Proceedings of International Specialist Seminar on Case Studies in Advanced Signal Processing, Peebles, 62–68.
3. CURTIS, T. E., and WICKENDEN, J. T., 1978, 'Digital Techniques for Beam Forming and Signal Processing', Digest of IEE colloquium on Modern Signal Processing Technology for Sonar, London, Digest 1978/52, 2/1–2/7.
4. DICKE, R. H., 1953, 'Object Detection System', US Pat No 2624876.
5. RABINER, L. R., SCHAFER, R. W., and RADER, C. M., 1969, 'The Chirp-Z-Transform Algorithm', *IEEE Trans*, **AU-17**, 86–92.
6. BUSS, D. D., VEENKANT, R. L., BRODERSEN, R. W., and HEWES, C. R., 1975, 'Comparison between the CCD CZT and the Digital FFT', Proceedings CCD 1975, San Diego.
7. WIDDOWSON, J. W., 1978, 'Charge-Coupled Device Implementation of the Chirp-Z-Transform', Digest of IEE colloquium on Modern Signal Processsing Technology for Sonar, London, Digest 1978/52, 10/1–10/5.
8. WIDROW, B., GLOVER, J. R., McCOOL, J. M., JAUNITZ, J., WILLIAMS, C. S., HEARN, R. H., ZEIDLER, J. R., DONG, E., and GOODLIN, R. C., 1975, 'Adaptive Noise Cancelling: Principles and Applications', *Proc. IEEE*, **63**, 1692–1716.
9. BOYLE, W. S., and SMITH, G. E., 1970, 'Charge-Coupled Semiconductor Devices', *Bell Syst Tech J*, **49**, 587–593.
10. DIX, J. F., DEAN, N., WIDDOWSON, J., and MAVOR, J., 1980, 'Applications of CCD's to Sonar Systems', *IEEE Proc*, **127**, Pt F, No 2, 125–131.
11. THOMSON, R., 1981, 'Time Delay and Integrate Device WM 2210', Wolfson Microelectronics Institute, Edinburgh, Document No 225/1415/RO2/81/11.
12. THOMSON, R., and DONALDSON, W., 1981, 'Tapped Delay Line WM 2020', Wolfson Microelectronics Institute, Edinburgh, Document No 212/1414/RO2/8119.
13. DENYER, P. B., McDOWELL, P. R., MILNE, A. D., REID, J. R. C., and THOMSON, R. S., 1982, 'Study of Enhanced Circuits for Beam-forming', Wolfson Microelectronics Institute, Edinburgh, Document No 285/1429/RO2/82/7, 52–77.
14. DEAN, N. H. E., DIX, J. F., and THOMSON, R. S., 1984, 'Sonar Beam-forming using Sampled Analogue CCD Delay Lines', Digest of IEE Colloquium on Signal Processing Arrays, London, Digest no 1884/25, 8/1–8/7.
15. DENYER, P. B., 1980, 'Design of Monolithic Programmable Transversal Filters using Charge-Coupled Device Technology', Thesis, University of Edinburgh.

16. DENYER, P. B., and MAVOR, J., 1978, '256 Point CCD Programmable Transversal Filter', Proc 5th International Conference Technology and Application of Charge-Coupled Devices, Edinburgh.
17. MILNE, A. D., McDOWELL, P. R., DENYER, P. B., REID, J. R. C., THOMSON, R. S., and McKEE, D., 1981, 'Development of CCD's for Sonar Signal Processing Final Report', Wolfson Microelectronics Institute, Edinburgh, Document No 231/1401/RO8/81/12.
18. DENYER, P. B., 1978, 'Interim Report on The Effect of Charge Transfer Efficiency on the Matched Filter Detection of Chirp Waveforms', Wolfson Microelectronics Institute, Edinburgh.
19. CLARKE, C., 1978, 'Sampling Considerations and FM Compression Systems', Digest of IEE Colloquium on Modern Signal Processing Technology for Sonar, London, Digest 1978/52, 9/1–9/4.

Chapter 11
Learning from biological echo ranging systems?

J. D. Pye
School of Biological Sciences, Queen Mary College, London, UK

Echolocation is a primary sense for orientation and navigation in bats, dolphins and some birds. A variety of highly effective systems has evolved to serve individual needs and it should be possible to learn much by studying their operation. But the task presents great problems both in biology and in physical analysis.

There are over 800 species of echolocating bats in 19 Families, ranging from carnivores and fish catchers to vampires, fruit eaters and nectar drinkers. Even insect catching is done in a variety of ways, each needing the appropriate instrumentation. Diversity is also increased because many insect prey operate acoustic countermeasures (ACM) so that the bats cannot optimise target detection but must use counter-countermeasures (ACCM) to optimise their energy capture to expenditure ratio.

Wideband signals may be brief impulses or deep f.m. sweeps, sometimes extending bandwidth by including harmonics. Long, narrowband Doppler signals are associated with fine tuning of emitter frequency, to compensate for the Doppler shifts of wanted echoes, and with synchronised ear movements that may give additional Doppler information. Because of strong atmospheric absorption at higher frequencies, long-distance detection needs low frequencies and also narrow bandwidth if correlation is not to be degraded. Bats therefore adopt different modes and stratagems and some vary their signals radically as range decreases during target interception.

Interpretation of signal properties is difficult because many of the premises of normal radar theory do not apply. Bandwidths often exceed an octave so that Doppler shift is not a simple offset or increment and one must use the true Doppler factor α. Flight speeds may reach 5% of the low propagation velocity in air, giving α values up to 1.1 and some bats may be able to distinguish between the accurate $(c + v/c - v)$ and approximate $(1 + 2v/c)$ Doppler formulae. Such high α values also produce appreciable temporal compression of the echo so that the "true" range of the target becomes relativistic and the skew of an ambiguity ridge for an f.m. sweep becomes arbitrary. Further, it does not seem that Doppler accuracy or the output of a Fourier transform receiver can be deduced from a $\tau - \alpha$ ambiguity diagram as is possible on the conventional $\tau - f$ axes of "narrowband" cases.

Wide bandwidths also lead to degradation of the correlation function with target velocity unless Doppler-tolerant hyperbolic f.m. (linear period rather than linear frequency) is used and many bats do this. But other f.m. patterns are found, harmonics are common and the spectrum may have a fine structure that results in prominent side-ridges on the ambiguity diagram. While some pulses can be assessed intuitively or by simple formulae, amplitude shaping may make it difficult to assign values even to such basic parameters as pulse duration or bandwidth.

The effort does seem to be worthwhile, however, for although some characteristics are likely to remain unique, others are already apparent to a lesser degree in underwater sonar and some will become relevant to radar with the advent of octave bandwidth signals. Nature is an experienced operator, for inadequacy leads to extinction, and many fundamental problems have been solved by the processes of evolution during the last 4.10^7 years.

Chapter 12

Acousto-optic correlators

R. Bowman*, P. V. Gatenby* and R. J. Sadler[†]

*Marconi Space & Defence Systems Ltd, The Grove, Warren Lane, Stanmore, UK
[†]GEC Research Laboratories, Hirst Research Centre, Wembley, UK

12.1 Introduction

Correlation is a fundamental signal processing operation which is extensively used for delay estimation, signal-to-noise enhancement, pulse compression and spectrum analysis in radar, lidar, sonar and communications systems. The most basic correlator comprises a mixer and an integrator, implemented in analogue or digital form, and forms the cross-correlation between a pair of signals for a single delay value. More advanced correlators may be assembled using tapped delay line devices which may be implemented using digital, c.c.d. or s.a.w. technologies. Such devices enable a discrete set of correlation values to be generated simultaneously over a range of delay values. The recent maturing of acousto-optic devices, silicon photodiode arrays and coherent solid-state light sources has given rise to new types of optical correlation devices which employ analogous tapped delay-line architectures. The attractions of the optical approach lie in the large instantaneous bandwidths (up to 1 GHz) or large range windows (up to 100 μs), the enormous time-bandwidth products (up to 10^7), the low output data rate, the programmability and flexibility, and the complex (phase coherent) operation which are permitted. In addition, optical systems are readily extendable to two-dimensional operation, enabling the simultaneous computation of two (orthogonal) parameters with even greater concurrency and flexibility. This chapter addresses the principles, architectures and performance of acousto-optic correlator systems. Two specific applications which might benefit from these devices are considered – namely, spread-spectrum signal processing and lidar ambiguity-function processing.

12.2 Optical components

The key components are the electrical-to-optical input transducers and complementary output transducers. The most versatile input transducer is the bulk

94 Acousto-optic correlators

acousto-optic Bragg cell. This solid state device comprises a piezoelectric transducer bonded to a block of optical material. The transducer converts the applied signal into an ultrasonic wave which gives rise to a spatially varying refractive index (through the photoelastic effect) in the material. A collimated and coherent optical beam incident at the Bragg angle will be diffracted by this "travelling phase grating". The Bragg cell has the following properties:

(i) A fraction of the incident light is deflected through an angle proportional to the r.f. frequency in the cell.
(ii) The optical frequency of the diffracted light is upshifted or downshifted by the r.f. frequency in the cell depending upon the angle of incidence at the cell.
(iii) The optical phase of the diffracted light follows the phase of the r.f. carrier in the cell.
(iv) At low signal levels the diffracted light amplitude follows the amplitude of the applied signal.
(v) Conjugate versions of the applied signal may be taken by using frequency upshifted and frequency downshifted diffracted orders.

Table 12.1 *Bragg cell materials*

Material	Acoustic velocity (mm μs^{-1})	Typical transit Time (μs)	Cell bandwidth
Lithium niobate	6.57	5	Large
Lead molybdate	3.63	5	Medium
Gallium phosphide	6.32	2	Large
Tellurium dioxide	0.62	50	Low
Dense flint glass	3.67	10	Low

Thus, the Bragg-cell acts as a continuously tapped delay line and (optical) single sideband mixer. Widely used Bragg-cell interaction materials are listed in Table 12.1. The transit time is the time a signal takes to traverse the optical aperture of the cell.

The output transducer may be realised in the form of a serial-output photodiode array. Such arrays comprise a linear or two-dimensional array of photodiode sensing elements defined on a silicon substrate. The photogenerated charge accumulated on each element is read out sequentially either by self-scanned m.o.s. shift registers or c.c.d. parallel in/serial out shift registers. Each sensing element provides an output proportional to the time-integral of the incident light intensity over the readout period of the array. Thus the array constitutes a bank of temporal integrators. Typical performance figures for linear and two-dimensional arrays are given in Table 12.2.

Acousto-optic correlators 95

In addition to these basic components, a light source, sundry spherical and cylindrical lenses, beam stops, spatial filters and polarisation filters are needed to complete a correlator system.

Table 12.2 *Detector array performance*

Parameter	Linear array	Two-dimensional array
Number of pixels	1024	385 × 576
Pixel dimensions	25 μm × 2.5 mm	22 μm × 11 μm
Clock rate	⩽ 5 MHz	7.4 MHz
Integration time	⩾ 0.2 ms	20 ms
Optical dynamic range (thermal noise limited)	40 dB	20 dB
Technology	self-scanned m.o.s.	frame-transfer c.c.d.
Reference	(1)	(2)

12.3 One-dimensional acousto-optic (a.o.) correlators

12.3.1 Optical architectures

Pre-1970, one-dimensional acousto-optic correlators invariably employed the spatially integrating approach in which a lens and a single photodetector integrate light diffracted from all points in the Bragg cell and the correlation appears in the temporal signal from the detector (Rhodes (3)). This approach suffers from a limited signal time window given by the Bragg-cell transit time and the need for a time-reversed version of the reference waveform. A more recent technique, introduced in the early 1970s, utilises the time-integrating property of photodiode arrays. The interferometric scheme shown in Fig. 12.1a illustrates the operating principles. Light from a c.w. laser is expanded to illuminate the aperture of the first Bragg cell at the Bragg angle (θ). A diffracted beam is produced, and this passes directly through the second cell since the Bragg condition is not satisfied and no further diffraction can take place. However, the angle of the second cell with respect to the undiffracted beam does satisfy the Bragg condition and a second diffracted beam is therefore produced at an angle of 4θ relative to that from the first cell. The two cells produce counter-propagating ultrasonic waves. The mid-plane of the cells is imaged onto the photodiode array and a spatial filter with two apertures in the Fourier-transform plane removes the zero-order light. The photodiode array is a square-law (intensity) detector and generates a term in the detected response given by:

$$\int_0^T S_1(t + \tau/2) \cdot S_2^*(t - \tau/2) \cdot dt + \text{bias} \qquad (1)$$

96 Acousto-optic correlators

Here the delay variable $\tau = 2x/v$, where the spatial coordinate x is measured from the cell centres and v is the acoustic velocity in the cells.

This is the symmetric cross-correlation of the signal waveforms $S_1(t)$ and $S_2(t)$ driving the two cells. The correlation is formed on a pedestal (bias) and modulates a spatial fringe pattern. The detector output is generally passed through a narrowband filter tuned to this carrier for bias removal and noise reduction. It should be emphasised that there is a wide diversity of time-integrating architectures, using different modes of a.o. modulation, different detection schemes, and different geometrical configurations. The performance of two time-integrating correlators realisable with currently available components is summarised in Table 12.3; one device is optimised for bandwidth, while the other is optimised for range window. A detector integration time of 10 ms is assumed.

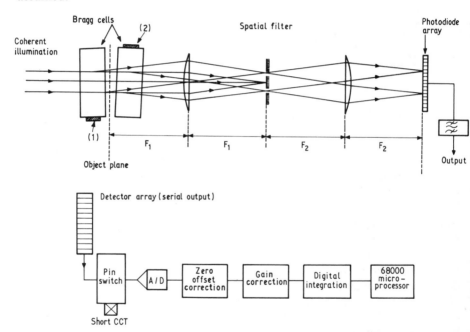

Fig. 12.1 *One-dimensional time-integrating correlator (a) Optical assembly. (b) Post detection processor*

12.3.2 Spread-spectrum signal processing

The large time-bandwidth products and range windows offered by time-integrating correlators make them exceptionally well suited to spread-spectrum signal processing. We confine the present discussion to direct sequence pseudo-noise codes which biphase $(0, \pi)$ modulate the transmitted carrier. In general, the received signals are buried in thermal noise and detection requires considerable signal-to-noise enhancement. A major problem associated with the

Acousto-optic correlators 97

reception of direct sequence data and communications transmissions is the initial synchronisation of the received code and the locally generated reference code. Conventional synchronisation algorithms are slow requiring a chip-by-chip search over the code phase uncertainty. The acousto-optic approach allows a simultaneous search to be made over the range window of the correlator. Thus, for a 10 Mbit PN code and 10 ms integration time, a conventional correlator would require 100 secs to correct a code misalignment of 1 ms. An a.o. time-integrating correlator with a range window of 100 μs could perform the same operation in 100 ms.

Table 12.3 *1D correlator performance*

Parameter	Correlator 1	Correlator 2
Bragg cell material	Lithium niobate	Tellurium dioxide
r.f. bandwidth	1 GHz	20 MHz
Range window	1 μs	100 μs
Time-bandwidth product	10^7	2×10^5
Dynamic range (fixed pattern noise removed)	45 dB	45 dB

The authors have constructed a number of a.o. correlators and assessed their performance over a period of several years. These systems have excellent prospects for miniaturisation and ruggedisation. Most of the limitations on performance may be traced to the detector array. Firstly, the array must adequately sample the output spatial fringe pattern. This demands more than 3 pixels per correlation response. Thus, in many cases the correlator range window will be determined not by the Bragg cell aperture but by the finite number of detectors in the array. Secondly, the dark current and gain of each pixel varies across the array in a random manner. In the presence of the pedestal (bias), the resulting fixed pattern noise can be 10% of the saturation level greatly exceeding all other noise sources. The most effective method of reducing this noise is to employ digital correction techniques. The block diagram for a post-detection processor is given in Fig. 12.1b. Following fixed pattern noise removal the authors have measured dynamic ranges greater than 45 dB. Thirdly, the detector determines the maximum integration period which can be obtained. Build up of charge due to dark current imposes an upper limit of several hundred milliseconds for a c.c.d. based device.

12.4 Two-dimensional correlators

12.4.1 *Architectures*
Two-dimensional acousto-optic correlators operate upon one-dimensional (temporal) signals but compute two parameters simultaneously on orthogonal optical axes. Such systems might employ acousto-optic devices which are 2-D

98 Acousto-optic correlators

in nature, for example, devices with very broad transducers or multiple-transducers. However, such devices are difficult to fabricate and rarely used. The majority of architectures employ 1-D a.o. devices. Both spatially integrating (s.i.) and time-integrating (t.i.) 2-D a.o. architectures have been reported. The latter may be employed for fine frequency resolution spectrum analysis using a 2-D chirp transform algorithm (Turpin 4)). The authors have constructed such a system with approximately 4000 resolved channels and 50 Hz resolution per channel. The 2-D t.i. system may also be used for radar ambiguity function (range-velocity) computation. In this case a 1-D chirp transform is performed on one axis of the system while a 1-D range correlation is performed on the other. The 2-D s.i. correlator is chiefly of interest for ambiguity function computation where relatively coarse Doppler frequency shifts are involved. The following section describes the construction and results obtained from such a system in the authors' laboratory.

12.4.2 Acousto-optic ambiguity function processor

A spatially integrating processor for realisation of the cross-ambiguity function may be constructed from a spherical and cylindrical lens as shown in Fig. 12.2a. (Said and Cooper (5)). Suppose the signals of interest $S_1(t)$ and $S_2(t)$ are real and represented by masks M_1 and M_2. The amplitude transmittance of the masks follows the signal amplitude in the t direction and is extended in the τ direction. Upon rotating M_1 counterclockwise about the origin through an angle θ, we generate the function $S_1(t\cos\theta + \tau\sin\theta)$. Similarly, for a clockwise rotation of M_2, we generate the function $S_2(t\cos\theta - \tau\sin\theta)$. Suppose now we modulate a coherent light beam with the product of the rotated masks and transform the resulting light distribution with the optical system shown in Fig. 12.2a. The light amplitude in the output plane will be proportional to:

$$\int_{-\infty}^{\infty} S_1(t\cos\theta + \tau\sin\theta) \cdot S_2(t\cos\theta - \tau\sin\theta) \cdot \exp(-2\pi jvt)\, dt$$

This represents a scaled version of the cross-ambiguity of the functions $S_1(t)$ and $S_2(t)$. In the special case when $\theta = 45°$, the light intensity detected in the output plane will be proportional to:

$$\left| \int_{\infty}^{\infty} S_1\left(\frac{t+\tau}{\sqrt{2}}\right) \cdot S_2\left(\frac{t-\tau}{\sqrt{2}}\right) \cdot \exp(-2\pi jvt) \cdot dt \right|^2$$

For real-time operation upon complex signal waveforms, the masks may be replaced by 1-D acousto-optic devices. Systems of this kind have been reported by Cohen (6) and by Guilfoyle (7). A breadboard system of this type has been assembled by the authors to assess the accuracy and identify the performance limitations which might be expected. The system is shown schematically in Fig. 12.2. A pair of orthogonal Bragg cells is driven by signal waveforms $S_1(t)$ and $S_2(t)$ respectively. Spherical and cylindrical lenses transform the beam to a horizontal line focus in the first cell, and a vertical line focus in the second cell.

The diffracted light amplitude arriving at some point (x, y) in plane P_1 is proportional to $S_1(t - x/v)$ and $S_2(t - y/v)$ where v is the acoustic velocity in the Bragg cells and x, y are the position coordinates in the horizontal and vertical cells. A dove prism rotated through an angle of $22\frac{1}{2}°$ to the vertical rotates plane P_1 through $45°$. Lenses L_1 and L_2 perform the desired spatial transform and imaging lens L_3 produces a magnified image at the output plane. Spatial filters are located at appropriate points in the optical train to remove

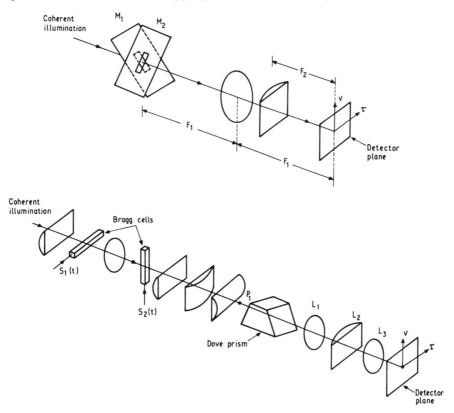

Fig. 12.2 *Spatially-integrating ambiguity function processors*

undiffracted light. Ideally, the laser light source is gated on while both signals $S_1(t)$ and $S_2(t)$ are completely contained within the aperture of the square mask. However, the system built by the authors employed a c.w. He–Ne laser and consequently, generated a time average (over the integration period of the camera) of the square modulus of the (truncated) ambiguity function. The performance of the system to date is summarised in Table 12.4.

The autoambiguity function of various waveforms produced by this system is shown in Fig. 12.3. The results are shown in an isometric format with

100 Acousto-optic correlators

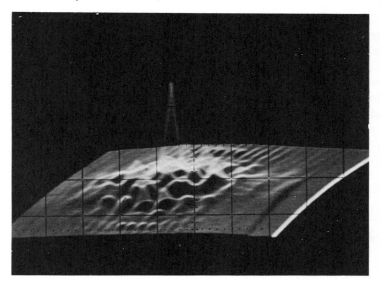

Fig. 12.3a *13 bit Barker code (2 Mbit clock rate)*

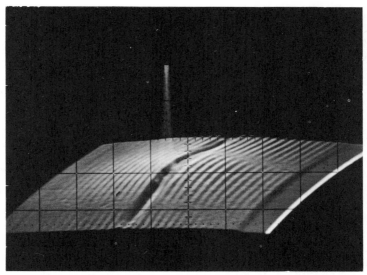

Fig. 12.3b. *Maximal length PN code (N = 20, 5 Mbit clock rate)*

frequency and delay lying in the horizontal plane. In all cases, the optically generated result was in very good agreement with the theoretical function. The chief performance limitation is the poor dynamic range (20 dB). This could be improved to approximately 40 dB by employing coherent (heterodyne) detec-

Acousto-optic correlators 101

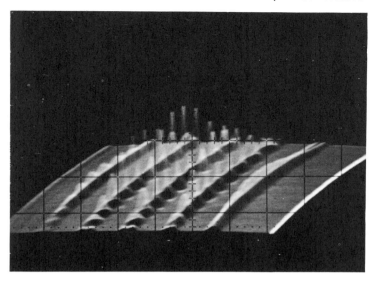

Fig. 12.3c *Rectangular pulse train – 200 ns pulse width, 1 MHz PRF*

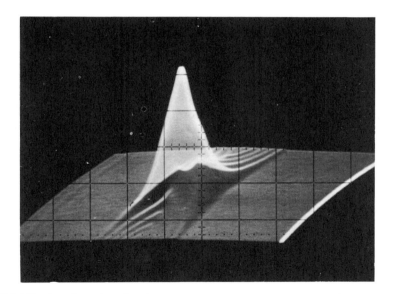

Fig. 12.3d *Single pulse – 1 µs duration*

Fig. 12.3 *Optically generated autoambiguity functions*

tion. However, implementation of this would be complex requiring provision of an optical distributed local oscillator. The dynamic range is limited by that of the 2-D detector (in this case a silicon vidicon). Consequently, 2-D arrays with improved dynamic range performance are required. Finer frequency resolutions, a greater number of frequency and range bins, and longer range windows could be obtained by using a pair of tellurium dioxide Bragg cells. Given improvements in dynamic range, this processor could be of some interest for the processing of coherent CO_2 lidar signals.

Table 12.4 *Ambiguity function processor performance*

Bragg cell material	dense flint glass
Signal time window	2.8 µs
Frequency resolution	560 kHz
Bandwidth	40 MHz
Number of frequency bins	71
Maximum range window	5.4 µs
Maximum number of range bins	216
Dynamic range	20 dB
Frame rate	50 Hz

12.5 Conclusions

Acousto-optic correlators with exceptional bandwidth or range window may be constructed using currently available optical components. Such devices should find application in radar, lidar or communications signal processing where a high degree of parallel processing is demanded. Two areas of particular interest are those of spread spectrum correlation and coherent lidar signal processing. The chief limitation on correlator performance comes from the relatively low dynamic range of the linear or two-dimensional detector array. Detector fixed pattern noise is the dominant noise source and digital post-detection circuitry is needed to correct for it.

12.6 References

1. Reticon data sheet, "S series solid state line scanners".
2. BURT, D. J., 1980, *The Radio and Elect Eng*, **50**, 205–212.
3. RHODES, W. T., 1981, *Proc IEEE*, **69**, 65–79.
4. TURPIN, T. M., 1981, *Proc IEEE*, **69**, 79–92.
5. SAID, R. A., and COOPER, D. C., 1973, *Proc IEE*, **120**, 423–428.
6. COHEN, J. D., 1982, *Proc SPIE*, **341**, 148–152.
7. GUILFOYLE, P. S., 1982, *Proc SPIE*, **341**, 199–208.

Chapter 13
Gaas MESFET comparators for high speed Analog-to-Digital conversion

L. Chainulu Upadhyayula
RCA Laboratories, Princeton, N.J., U.S.A.

13.1 Introduction

A full parallel comparator (flash converter) architecture is employed in fast Analog-to-Digital converters. The input is applied to all the comparators simultaneously. The response time is the time required for any one comparator to make a decision and produce an appropriate output. The comparator is thus the critical element that determines the ADC speed. Silicon bipolar differential amplifier type comparators have been commonly used in fast ADCs. In a bipolar transistor, a small change in base-emitter voltage (V_{BE}) of about 0.1–0.2 V ($\approx 4.4\,kT/q$) brings about an order of magnitude change in the collector current which produces a large voltage swing across the load resistor. Furthermore, the small V_{BE} change required to switch the transistor from full ON to the OFF state makes it possible to readily introduce reference voltages for the comparators. The fastest silicon ADC operating at 300–400 MHz was reported by De Graaf (1). The power dissipation, however, becomes excessive and current gain falls-off as the operating frequency is increased above this value. Advanced signal processing systems presently under development and future systems will require GHz sampling rate ADCs. It is, therefore, necessary to study other devices with superior performance. GaAs MESFET exhibited a gain-bandwidth product of 15 GHz. The work of Liechti (2) and Long et al. (3) demonstrated that GaAs MESFET is suitable for analogue and digital applications at microwave frequencies. Following silicon bipolar-type ADC development Curtice (4) and Greiling (5) studied GaAs FET differential amplifier type comparators. These computer simulations predicted substantial change in threshold levels can occur due to variations in material and device parameters (4). The differential amplifier gain is at best 2–3 (5) for linear loads. The problem of setting the threshold levels was not addressed in either of these studies. Furthermore, no experimental results have been reported on GaAs FET dif-

104 GaAs MESFET comparators

ferential amplifier type comparators. The small current gain and the difficulty in establishing reference voltages in field effect transistors (FETs) may be responsible for the lack of progress. Meignant and Binet (6) recently reported on GaAs FET strobed comparators. The results show that output transition from LOW to HIGH is very slow compared to the one from HIGH to LOW indicating that the conversion takes several regeneration time constants of the comparator before a correct decision is made. Also, the response of these strobed comparators for a sample-and-hold type input was not studied. This leaves us with the choice of studying other circuit approaches to implement the high speed comparators with GaAs FETs. The author (7, 8) has proposed a novel GaAs MESFET comparator for GHz sampling rate operation with 3- to 4-bit resolution. This paper will review the principle of operation, design procedures, and describe the fabrication and performance of 2- and 3-bit ADCs.

13.2 Principle of GaAs FET comparator

The output of a comparator circuit remains constant until the input signal reaches some reference or threshold level. The over-drive voltage (ΔV_i) required above the reference level to change the output voltage from logic LOW to HIGH or vice-versa determines ADC resolution. It was pointed out earlier that it is not readily feasible to introduce reference voltage levels in FET circuits. It is therefore necessary to build these threshold levels as an integral part of the comparator. The proposed GaAs FET comparator (Fig. 13.1a) consists of a driver or switch FET (T_s), a constant current load FET (T_L) and capacitive load (C) at the output node. The drain saturation current for the switch FET is considerably higher than that for the load FET (Fig. 13.1b.). The comparator output (i.e. the voltage at the drain node of the switch FET) is level shifted in a conventional circuit to make it compatible with the input voltage levels required for driving the depeletion-mode FET logic.

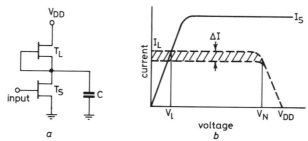

Fig. 13.1. (a) A novel GaAs MESFET comparator. (b) Current-voltage characteristics of MESFETs used in the comparator. (After Upadhyayula Ref. 8)

The comparator output makes a transition (from LOW to HIGH or vice-versa) when the currents in the switch and load FETs are equal. The currents are equal when the negative input voltage at the gate of the switch FET reaches the

GaAs MESFET comparators

appropriate threshold level. Therefore, the threshold of the GaAs FET comparator is given by the relation:

$$I_{DS}/I_{DL} = K(1 - \eta^{0.5})/(1 - \eta_s^{0.5}) = 1.0 \tag{13.1}$$

where $\eta_s = V_B/V_p$, $\eta = (V_{th} + V_B)/V_p$ and $K = Ws/W_L$. V_B, V_p, I_D and W are Schottky barrier voltage, channel pinch-off voltage, drain current, and device width, respectively. The above equation can also be written in the form:

$$1/K = (1 - \eta^{0.5})/(1 - \eta_s^{0.5}) \tag{13.2}$$

K or $1/K$ is used as a design parameter. The Schottky barrier voltage is relatively constant for GaAs for a given process and ranges from 0.4–0.6 V. For any desired threshold levels (V_{th}), the parameter K can be computed from the above expression for a fixed pinch-off voltage. For a set of threshold levels we obtain a corresponding set of K values. It is important to note that the threshold levels depend only on the ratio of the widths of the switch and load FETs and not on their absolute values.

The FET widths are fixed by the over-drive voltage and response time requirements. For input signal levels less than or equal to the threshold level, the drain voltage of the switch transistor is low (V_L). When the input signal level exceeds the threshold level by ΔV_i the drain voltage switches to V_H. The capacitive load presented at the drain must be charged or (discharged) for this voltage transition to occur. The rate at which the drain voltage changes is given by:

$$\Delta t = C \cdot \Delta V/\Delta I \tag{13.3}$$

where ΔI is the current available for charging (or discharging), C is the load capacitance, ΔV is the voltage transition ($V_H - V_L$), and Δt is the switching (rise) time. In our GaAs FET circuits, C and ΔV are of the order of 0.02 pF and 3 V, repectively. For the switching time to be about 100 ps, a current change (ΔI) of 0.6 mA is needed. Also, ΔI should be small compared to I_S (Fig. 13.1b) so that several comparator threshold levels can be set between I_S and zero current levels. The change in FET drain current is caused by a corresponding voltage change ($\Delta Vg = \Delta V_i$) at the gate and is given by

$$\Delta I = g_m \cdot \Delta V_i \tag{13.4}$$

For the comparator to operate satisfactorily, the over-drive voltage (ΔV_i) should be less than or equal to 1/2 of the quantization level (Q). If $Q \cong 0.2$–0.4 V, the maximum value for ΔV_i is 0.1 V. An excellent FET performance is obtained with channel doping of about $1.0 \times 10^{17} cm^{-3}$. As shown by Fair (9), a 100 μm wide GaAs FET with 5–6 V pinch-off voltage will have a g_m of 7.0–7.5 ms. The minimum size of the switch FET as fixed by the over-drive requirement is 100 μm. The widths of the load transistors are then obtained from the $1/K$ values determined earlier.

106 GaAs MESFET comparators

13.2.1 Comparator design

The highest speed of operation is achieved in flash (parallel comparator) ADC. The number of comparators required is $2^n - 1$. Two and three bit ADCs therefore require 3 and 7 comparators, respectively. Assuming a full scale-voltage of 1.6 V, the corresponding quantization steps are 0.4 and 0.2 V. Comparators are then designed with the help of the equations developed in the previous section and the device parameters are obtained (Table 13.1). The Schottky barrier voltage (V_B) and the channel pinch-off voltage (V_p) are used as parameters in our design.

Table 13.1 Comparator design parameters for a 2- and 3-Bit ADC

ADC	Comparator Threshold Voltage (V)	$V_B = 0.4$ V		$V_B = 0.6$ V	
		$V_p = 5.0$ V	$V_p = 6.0$ V	$V_p = 5$ V	$V_p = 6$ V
2-Bit	0.4	0.837	0.856	0.845	0.865
	0.8	0.711	0.745	0.720	0.756
	1.2	0.606	0.652	0.612	0.661
3-Bit	0.2	0.911	0.922	0.918	0.929
	0.4	0.837	0.856	0.846	0.865
	0.6	0.771	0.798	0.781	0.808
	0.8	0.711	0.745	0.720	0.756
	1.0	0.657	0.697	0.665	0.707
	1.2	0.606	0.652	0.612	0.661
	1.4	0.558	0.610	0.562	0.618
	1.6	0.513	0.570	0.515	0.577

13.2.2 Decoder design

Comparator outputs must be converted to provide bit outputs. The bit outputs may be in either binary or Gray code format. The complementary outputs are not available in MESFET comparators and must be generated using separate inverters. Simple 2–4 input NAND/NOR gates can be used to implement the decoding logic.

Buffered FET logic (BFL) is selected for the circuit implementation because of its versatility and higher speed. To minimize power dissipation, a channel pinch-off voltage of about 2.0 V is used for the logic circuits.

13.2.3 Computer simulations

An R-CAP computer simulation program with a user-defined GaAs FET model of reference (10) was used to study circuit performance. Device parameters for the model were obtained from the measurements on discrete test FETs. Simulated response of the comparators for 2-bit ADC is shown in Fig. 13.2. The threshold levels agree with the design values and the over-drive voltage is 0.1 V.

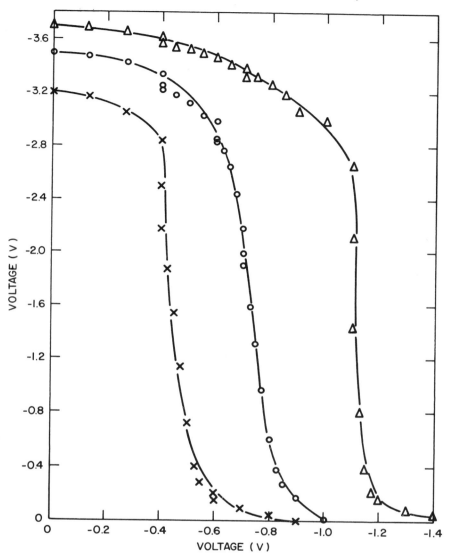

Fig. 13.2. *Computer simulation of the threshold characteristics of comparators designed for 2-bit ADC. (After Upadhyayula et al., Ref. 11.)*

The response time for the comparator and coding logic together is about 400–600 ps. These results assure 3-bit resolution at 1.0 GHz sampling-rate operation. A two-stage comparator will improve the sensitivity and make a 4-bit ADC design feasible.

13.3 IC fabrication

Two- and three-bit ADCs were fabricated using standard optical photolighography and contact printing techniques. GaAs epitaxial wafers grown by vapour-hydride technique were used. The nominal doping and thickness of these wafers are $8-10 \times 10^{16} \text{cm}^{-3}$ and $0.6\,\mu\text{m}$, respectively. The wafers were electronically thinned to the limit to obtain uniform device characteristic across the wafer. Seven masking steps were required. Au:Ge/Ni/Au ohmic contacts, Ti/Pd/Au Schottky barriers, Ti/Au interconnections and polyimide isolation were used. Because of the different pinch-off voltages used, the comparator and logic sections were processed separately.

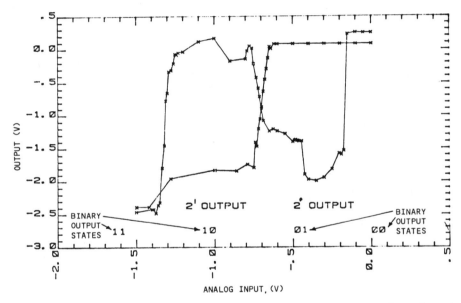

Fig. 13.3. *Measured response of a 2-bit ADC to dc input voltage. (After Upadhyayula et al., Ref. 11.)*

13.4 Experimental results

The ADCs fabricated consisted of comparators and associated coding logic circuits. Sample-and-hold and digital-to-analog circuits were not included. A 0 to 1.6 V variable input was fed to the comparators and the corresponding binary or Gray code outputs were monitored on a sampling oscilloscope. The input pulse width was varied to determine the response time and the pulse repetition rate was varied to determine the effective sample rate. The response of a 2-bit ADC to a dc input is shown in Fig. 13.3. Comparator thresholding and output encoding are clearly seen. The measured quantization step is 0.55 V, slightly larger than the design value (0.4 V). This difference between the design and experimental values can be explained in terms of the variation in FET pinch-off voltage. Also, an off-set error of 0.2–0.4 V was observed. The IC was operated with 0.5 to 0.7 ns wide input pulses and repetition rates as high as 1.0 GHz. The three-bit ADC was also satisfactorily tested for GHz sampling rate operation. Details of the performance of 2- and 3-bit ADCs including the gain and off-set errors can be found in reference (11).

13.5 Conclusions

A novel GaAs MESFET comparator for high speed and medium resolution ADC has been described. These comparators do not require any reference voltages. Threshold levels are built into comparators themselves. The comparators and coding logic for flash converter type 2- and 3-bit ADC were designed and fabricated in monolithic form. The ICs were operated at an effective sampling rate of 1.0 GHz. A two-stage comparator design will improve the sensitivity and provide 4-bit ADC resolution. Sample-and-hold circuits are being developed at various laboratories. Master-slave flip-flops can be used for latches. Therefore, a single-chip monolithic 3- or 4-bit GaAs FET ADC for gigahertz sampling rate operation can be fabricated in the near future. Further work is required to study the errors and temperature effects before incorporating these ADCs into real systems.

13.6 References

1. DEGRAFF, K., 1978, 'A Silicon 400MS/S 5-Bit A/D Converter', Workshop on High Speed A/D Conversion, Portland, OR, U.S.A.
2. LIECHTI, C. A., 1976, 'Microwave Field-Effect Transistors', *IEEE Trans. Microwave Theory Tech.*, **MTT-24**, 279–299.
3. LONG, S. I. *et al.*, 1982, 'High-Speed GaAs Integrated Circuits', *Pro. IEEE*, **70**, 35–45.
4. CURTICE, W. R., Private Communication.
5. GREILING, P. T., Private Communication.
6. MEIGNANT, D., and BINET, M., 1983 GaAs IC Symposium, Phoenix, AZ, U.S.A.
7. UPADHYAYULA, L. C., 1979, 'GaAs FET Comparators for High-Speed Analog-to-Digital Conversion', GaAs IC Symposium, Lake Tahoe, NE, U.S.A.
8. UPADHYAYULA, L. C., 1980, 'MESFET Comparators for Gigabit-Rate Analog-to-Digital Converters', *RCA Rev.*, **41**, 198–212.
9. FAIR, R. B., 1974, 'Graphical Design and Iterative Analysis of the dc Parameters of GaAs FETs', *IEEE Trans. Electron Devices*, **ED-21**, 357.
10. CURTICE, W. R., 1980, 'FET Model for Use in the Design of GaAs Integrated Circuits', *IEEE Trans. Microwave Theory Tech.*, **SMTT-28**, 148.
11. UPADHYAYULA, L. C., *et al.*, 1983, 'Design, Fabrication and Evaluation of 2- and 3-Bit GaAs MESFET Analog-to-Digital Converter IC's, *IEEE Trans. on MTT*, **MTT-31**, 2.

Chapter 14

Designing in silicon

Douglas Lewin
University of East Anglia

14.1 Introduction

This paper provides an overview of the various design methods available for the realisation of digital logic circuits used in signal processing with particular emphasis on silicon structures.

The traditional methods of designing logic systems by deriving an optimum interconnection of basic logic modules (such as the SN74 series) using a mix of intuitive and theoretical techniques are no longer applicable to today's technology. The major reason for this lies in the complexity of the systems which we are now capable of designing and the need to realise these designs in terms of integrated circuit technology – *designing in silicon*.

This new technology necessitates major changes in design methods, based on the use of computer-aided design techniques (1). Software tools are now essential to the design process enabling the engineer to model and evaluate his designs using simulation and analysis programs, to determine and specify test requirements and to perform efficiently the layout and interconnection functions. An important design decision is the type of technology chosen for the realisation: Table 14.1 shows the methods currently available. The Table is of necessity greatly simplified and any final design decision must be determined by the actual system requirements, the available resources and the possible trade-off's between alternative approaches such as hardware/software realisation. For example, for a microprocessor system program development time is a major consideration whereas in a LSI/VLSI design the major factor is the layout and testing of the chip. In both cases specialist software tools would be required to facilitate development. However, if flexibility was a prime design factor microprocessor realisation might be chosen. On the other hand if security of design (in the sense of plagiarism by other manufacturers) was a major consideration then an LSI implementation would be preferred.

In signal processing an important consideration is speed of operation and here the microprocessor realisation would be far too slow for many real-time applications. In this case a hardware implementation is inevitable and the use

Designing in silicon 111

of LSI/VLSI techniques opens up the possibility of realising complex processing algorithms directly in terms of silicon (2). When realising logic systems in terms of LSI circuits both *semi-custom* and *full-custom* design is possible.

Table 14.1 *Comparison of available technology*

Parameter	Custom LSI	Semi-Custom LSI	Discrete MSI/SSI	Microcomputers
Speed	Fastest	Fast	Medium	Slow
Size	Smallest (chip size)	Small (chip size)	Medium (many chips)	Medium (many chips)
Development Time	Longest (layout)	Long (layout)	Medium (logic design)	Short-medium (software)
Flexibility	Lowest	Low	Medium	High
Initial 'Tooling' Investment	Highest (layout Test)	High (layout Test)	Medium (Logic design)	Low Medium (Software)
Unit Costs:				
High Volume	Lowest	Low	Medium	Highest
Low Volume	Highest	High	Medium	Lowest
Reliability	Highest	High	Low	Medium
Security	Highest	High	Low	Medium

14.2 Semi-custom IC design

Semi-custom design isolates the designer from much of the complexity of chip design since the majority of the processing steps are completed before committing the device to a particular function. The designer, working with pre-defined components or logic cells, is concerned primarily with producing an interconnection layout.

The simplest form of semi-custom technology is the programmable *ROM and PLA* which utilise either minterm or product terms in an AND/OR structure. These devices are programmed either by defining the mask patterns used for the final metalization stage of the fabrication process or by direct field programming using fusable links or avalanche techniques (3) (4).

Systems using these devices can be developed using conventional logic design skills but they are limited in the range of application. PLA designs in particular are constrained by the number of available gates on the chip (in the order of a few hundred). ROM's though widely used, especially to provide hardware realisation of software, suffer from the disadvantage of being easy to copy and hence are insecure.

By far the most versatile of semi-custom LSI technology is the gate array, or uncommitted logic array (ULA). A gate array is an LSI chip on which gates, or individual components of gates such as enhancement or depletion transistors, are placed in a regular matrix format with the gate (component) interconnections still to be completed. The chips are preprocessed up to the final stages of metal layer patterning (which defines the connections and contacts). Thus only one or two metalising masks have to be custom designed and processed instead of the full mask set required for complete custom design (at least 6 masks). Gate arrays are available in most semi-conductor technologies giving the designer the freedom to choose the most appropriate device in terms of number of gates, power and switching times etc. For example:

CMOS – order of 5,000 gates; low power; approx. 2–3 ns
TTL – order of 1,200 gates;   medium power; approx. 2–3 ns
ECL – order of 1,000 gates;   high power; approx. 1 ns
I^2L – order of 4,000 gates; medium power; approx. 6 ns
GaAs – order of 1,000 gates; low power; approx. 200 ps

There are numerous manufacturers of gate arrays and new products appear regularly; the maximum number of gates is in the order of 10,000.

The layout and interconnection of gates can be performed manually using special layout grids but CAD tools are essential for all but the smallest designs. Manufacturers also provide libraries of standard layouts for the basic logic configurations often available as decals for use on manual layout grids.

An alternative technique is the *standard cell* approach in which compact layouts for basic gates and circuits (such as those available in the SN74 series logic) are generated by expert designers and stored in a computer *cell library*. When every cell has the same height, though its width may differ, the technique is known as *polycell design*. The design process is analogous to conventional logic design using MSI/LSI modules; appropriate standard cells are selected from the library and placed in rows across the chip, leaving sufficient space for interconnections. The process is obviously highly dependent on computer-aided design tools, requiring at least graphics and placement/interconnection aids. Note that the technique is in effect a constrained version of a full custom design and consequently the complete set of fabrication masks must be generated. The initial investment costs are much higher than gate arrays and the technique would only be viable for large production runs of more than 100,000 chips.

Designing in silicon 113

The user can either provide the supplier with a logic design and corresponding test pattern data (using a test description language) or a full chip layout; a design procedure for realising gate array systems would be as follows:

System Design
Specification of system at behavioural level. Partitioning of system into functional modules realisable as gate array chips with minimum interconnections.

Logic Design
Using normal logic design practice produce designs for gate arrays; estimate array size in terms of input, output, power pins and number of equivalent gates. Evaluate designs using functional and logic simulators. Generate test sequences.

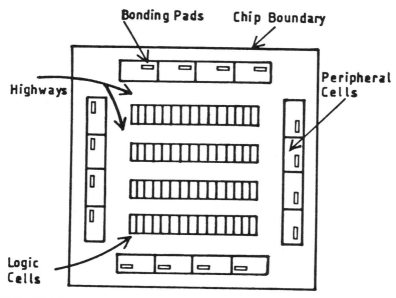

Fig. 14.1 *ULA layout*

Physical Design
Translate logic designs into array cell form using standard configuration libraries if available; a typical ULA layout is shown in Fig. 14.1. This is the layout and interconnection stage. Finally the layout must be checked to ensure that the design rules appropriate to the technology have not been violated; an analysis of the electrical characteristics of the actual circuits could also be required at this stage.

The advantages of ULA's are that:

● With standard cell libraries only a small number of cells need to be carefully designed and laid out.

114 Designing in silicon

- CAD programmes are available for layout and routing.
- Cost effective in very low production volumes (100's)
- Designs are as difficult to copy as custom LSI.
- Wide choice of technology is available.
- Arrays are available which allow both analogue and digital functions to be integrated on the same chip.

but there are disadvantages:

- Gate array chip size at least double that of custom LSI because of the need to reserve space for interconnections and the high percentage of unused gates.
- It is difficult to keep gate delays uniform thereby increasing risk of circuit hazards.

14.3 Custom LSI design

In this case the same basic top-down design approach still applies but partitioning is effected on the chip itself, attempting to produce a regularly structured architecture with the objective of building up a system by replicating basic cellular structures. Note that the achievement of optimum surface area is more dependent on the 'wires' and overall circuit topology than the actual number of gates. This necessitates relating the functional partitioning to actual chip areas, the 'real-estate', at the lowest level. As with semi-custom design simulation tools, at both gate and functional levels, are required to check out designs and generate test sequences. In addition it is necessary to perform small signal analysis of the MOS circuits to determine circuit delays, timing errors, etc.

One important exception however is that if conventional logic design techniques are used with subsequent translation into, say, NMOS transistor circuits, then it is very difficult to optimize areas on the chip, and more important, serious errors can arise in the translation process. This has led to design methods which allow the actual geometrical layout of the logic circuits to be reflected in the logic diagrams. One such method is the 'Sticks Diagram' (5) which represents the polysilicon, diffusion and metal layers using the colours, red, green and blue respectively; contacts are represented in black: the representation is shown in Figure 14.2(a). Another important consideration is to ensure that the design rules for a particular technology specifying allowable values (usually in microns) for widths, separations, overlaps, extensions, etc. for the geometric patterns can easily be applied at the physical design stage. Mead and Conway (6) have parametised these values using a common metric λ based on typical mask tolerances and process deviations to express the permissible misalignment between levels. Using generalised λ-based design rules LSI circuits can be reliably and simply designed. However, the technique leads to a conservatively toleranced design and if the system requires higher performance it will be necessary to exploit the technology further.

Designing in silicon 115

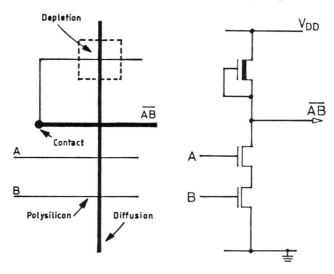

a) Sticks Diagram

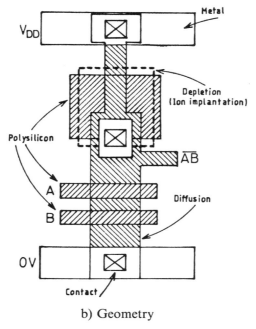

b) Geometry

Fig. 14.2 *Physical design of LSI NAND*

14.4 Computer aided design tools

CAD tools for the system and logic design stages are basically the same for both semi- and custom LSI and consist of software packages for logic simulation at gate and register/highway levels, timing analysis and fault test generation. Though placement, routing and interconnections can be done manually for semi-custom circuits with less than 500 gates in general software tools will be required for these functions.

For the physical design of custom LSI/VLSI (7) the tools will be similar but more sophisticated falling into three main categories:

(a) layout and interconnections
(b) layout verification and design rule checking
(c) functional and parametric testing

IC layout is the function of translating the logic specification for a chip into a set of photo-lithographic masks used in the fabrication process. Mask geometries are comprised almost entirely of rectangular shapes and wires (see Figure 14.2(b)) and to handle the attendant complexity CAD methods are essential. The initial layout is inputted using digitisation, high-level geometric programming languages or interactive graphics based on sticks diagrams or geometric shapes. After compactation, layout and interconnection routines have been applied the layout description would be stored in a design file and used as a output to plotter's, VDU's and mask cutting machines. Mask designs must be checked in two ways, firstly to ensure that the design rules have not been broken and secondly to verify that the physical circuit still meets the logic specification. After fabrication the logic functions (as determined by the test routine) and process parameters (circuit delays etc.) of the chip must be tested.

It is apparent that as the complexity of gate arrays increases, up to a maximum of 10,000, there will be little difference in the CAD tools required for either custom or semi-custom design.

There are many problems still to be solved in the CAD of VLSI circuits in particular the specification and evaluation at the behavioural level, the generation of test procedures, and simulation at the MOS level. One important development is the use of silicon compilers which allow a textural description of the logic/system requirements to be used as the input to an automatic mask-generation programme. This has already been successfully achieved for PLA structures and gate arrays.

14.5 The users point of view

The decision to use IC technology is not an easy one to make and many factors have to be considered, for example:

- Motives – product gains, speed/cost/reliability/size, etc. new design or replacement, design modification expected, security of design

- Supplier – technology/performance/support/CAD available, etc.
- Design automation – in-house or suppliers
- Learning curve – can take 6–18 months to master system

Full-Custom Design

- Essential to use in-house CAD system.
- Large initial investment in staff and equipment
- Need to find a "silicon foundry" to fabricate chips
- Investment for future – Semi-Custom LSI will lose effectiveness in order of 10,000 gates
- Considerable development remains to be done on producing effective CAD work stations
- Design process can be very staff intensive (5 man-years for processor chip)

It is apparent that there are many problems to be overcome in designing in silicon, especially if the custom route is chosen. Nevertheless, it is inevitable that this will become the normal method for realising logic systems. Moreover if the VLSI technology is fully exploited by using highly parallel structures, then there are incomparable advantages to be gained.

14.6 References

1. RAYMOND, T. C., 1981, 'LSI/VLSI Design Automation', *IEEE Computer*, **14** No. 7, 89–101.
2. KUNG, H. T., 1979, 'Let's Design Algorithms for VLSI Systems", *Caltech Conf. on VLSI Jan. 1979*, 65–90.
3. UIMARI, D. C. 1970, 'Field Programmable Read-Only Memories and Applications', *Computer Design*, **9**, Dec. 49–54.
4. MITCHELL, T. 1976, 'Programmable Logic Arrays', *Electronic Design*, **19**, No. 15, 95–101.
5. WILLIAMS, J. D., 1977, 'STICKS – A new approach to LSI Design', *MSc Thesis M.I.T.* June.
6. MEAD, C. and CONWAY, L., 1979, 'Introduction to VLSI Systems', *Addison-Wesley New York*.
7. TRIMBERGER, S., 1982, 'Automating Chip Layout' *IEEE Spectrum* June 38–45.
8. WALLICH, P., 1984, 'On the Horizon: fast chips quickly' *IEEE Spectrum*, March 28–34.

Chapter 15

Very high performance integrated circuits

A. L. Mears
Royal Signals and Radar Establishment, Malvern, UK

15.1 Introduction

One of the most significant factors in future military and civil electronics will be the use of very high performance integrated circuits (VHPIC). In military systems, VHPIC will provide data throughput rates two orders of magnitude superior to current capability, in applications such as smart weapons, multi-role radars, sonars, communications, command and control, and electronic warfare. VHPIC application specific circuits will increasingly dominate design, embodying the key features particular to individual systems and equipments. It will therefore become vital for companies selling electronic systems to have adequate access to advanced silicon VLSI technology and CAD.

15.1.1 *UK and US programmes*
Because of the importance of VHPIC for Defence, both the US and UK military are funding major research programmes. The US Very High Speed Integrated Circuit Programme (VHSIC) is reputed to be DOD's highest priority technology programme [1] [2]. In the UK, the funding of a similar (though smaller) 5-year VHPIC programme started in 1983. Both the DOD VHSIC Programme and the MOD VHPIC Programme have adopted a dual technology/applications approach, funding advanced VLSI technology research and in parallel funding 'brassboards' or 'demonstrators' to develop the expertise in designing and exploiting VHPIC processors and to provide an applications-driven lead for the technology work. A recent third major emphasis in the US VHSIC programme is provision of 'insertion' funding to assist 15 existing system programmes to pursue a VHSIC approach in parallel with their more conventional developments in order to introduce VHSIC into service as rapidly as possible [1].

15.1.1.1 *Alvey technology programme*
In the UK, VHPIC technology research is funded through the joint Industry-DTI-MOD-DES Alvey programme [3]. The fact that MOD decided to link its

VHPIC technology programme into Alvey reflects an appreciation, as discussed later, that there is a strong commonality between military and civil applications of VHPIC. VHPIC and information technology generally are areas where the military and civil industries can gain substantially from collaboration.

15.1.1.2 Alvey and VHPIC application demonstrator projects

The VHPIC Application Demonstrator Programme (VAD) is funded by MOD and industry, outside the formal Alvey Programme but co-ordinated with it. VAD is concerned with exploitation of VHPIC for military applications requir-

Table 15.1 VHPIC application demonstrator phase 1 contracts

Topic	Contractor
Distributed Array Processor	ICL
Array Processor	Ferranti Computer Systems
Reconfigurable Signal Processor	Marconi Avionics
General Purpose Sonar Processor	Plessey Marine
Multipurpose Systolic Array Node Chip	STL
Two Dimensional Imaging and Tracking	Marconi Defence Systems
General Purpose Chip Set for Weapons	Ferranti Instrumentation
Radar Plot Extractor Processor Chips	Ferranti Computer Systems
ESM Signal Processor	Racal Electronics
Universal Tactical Radar	Plessey Electronic Systems
Zero IF Digital Radio Receiver	STL
Digital Map and Raster Generator	Marconi Avionics
Direct Voice Input	Marconi Avionics

ing very high throughput digital signal processing. The Alvey Programme also includes demonstrator projects and these emphasise information technology and intelligent knowledge based systems (IKBS), and are therefore complementary to the VAD projects. In addition there are 'exemplar' projects to demonstrate techniques in particular limited domains. All of these applications projects will use advanced VLSI processing and CAD, and a major driving force in the Alvey VLSI-CAD Programme is therefore the need to provide advanced technology on a timescale matched to the applications projects and to meet the urgent needs of UK systems companies for high performance VLSI technology.

15.1.1.3 VHPIC application demonstrator programme

Over 50 proposals were received from industry for the VAD phase 1 studies, but within the available funding it was only possible to pursue 13 of these (Table 15.1). The phase 2 VAD Projects are expected to start in early 1985.

120 Very high performance integrated circuits

15.2 VHPIC process technology

In the early 1960's integrated circuits contained only a few transistors, and each transistor was more than 100 times bigger in area than the transistors in present day ICs. In the past 20 years chip sizes have increased to around 10 mm square and more efficient circuit concepts have been introduced. Coupled with the reduction in transistor size this now enables processors to be made with typically 50,000 transistors per chip, and some research processors have been produced in the US and Japan with as many as a million transistors per chip.

15.2.1. *Alvey-funded process research*

MOD has funded industry research on 3 major silicon techologies: CMOS-SOS/CMOS-SOI at GEC and Plessey, ECL at Plessey and CDI at Ferranti. The key factors in selecting these three technologies have been nuclear hardness, suitability for high yield commercially viable processes and scope for VLSI-scale integration and very high data throughput. Funding of these technologies is now handled via the Alvey VLSI programme, but with an additional major emphasis on bulk CMOS. Bulk CMOS is seen as the most important technology for civil applications, but its military role is uncertain because of its nuclear vulnerability. Nuclear hardening of VHPIC technology is discussed by Mears [4].

15.2.2 *VHPIC process performance – Gate-Hz/Watt*

At the process level first generation VHPIC circuits will have a minimum transistor feature size of about 1.25 microns. By comparison the geometry of UK IC processing currently in production is 2 – 5 microns. Smaller transistor size improves circuit performance not only through higher circuit complexity, but also by increasing circuit speed and reducing power dissipation per transistor. VHPIC chip complexity will be several hundred thousand transistors and the technology must provide the capability to exploit efficiently this complexity and the increased speed. This implies at least 2 levels of metal interconnect: otherwise the interconnection paths become so tortuous that most of the chip is wasted in interconnect and the speed is also degraded. Large on-chip memory may also be required to minimise global communication.

For CDI and MOS technology a minimum feature size of 1.25 microns translates into processing performance (functional throughput rate) of about 3×10^{12} gate-Hz/W. ECL is a factor 5 lower than this because of its high power consumption. Functional throughput rate (FTR) is a useful measure of technology performance but must be treated with caution because it depends on circuit detail as well as raw speed. Gate speed at 1.25 microns for CDI and MOS is in the range 250 – 500 MHz which would give a FTR of about 2×10^{13} gate-Hz/W for a 60,000 gate chip. In practice for complex circuits, factors such as local memory access time propagation delay, clock skew and sub-circuit cycle times reduce the clock rate by typically a factor of 8 giving the FTR of 3×10^{12} gate-Hz/W quoted earlier. But, for circuits such as systolic arrays, where data flow is from

Very high performance integrated circuits 121

element to adjacent element and memory access delay is minimised, the chip can in principle be clocked at close to its raw gate speed. Therefore one must exercise care if using FTR itself as a definition of VHPIC perfomance. Also one should note that past improvements in IC performance have come not only from reduction in geometry and increase in chip area, but also from innovations in circuit design such as efficient RAM cells, and there may be further important improvements to be achieved from new circuit element concepts.

15.2.3 *VHPIC system performance*

At the systems level the conventional measure of performance is the computation throughput, measured in millions of operations per second (MOPS). One operation is defined as a 16 × 16 bit real fixed point multiplication, which is a pertinent definition for VHPIC applications where usually performance is dominated by the time required for multiplication.

The system performance in MOPS depends on the FTR, but also on the architecture, programmability and choice of algorithm, which can have a quite subtle but major impact on system performance. Performance may be lost through conflict for resources, bus contention, access time to external memory or poor matching of the computation to the available parallelism of the processor. At one extreme a single function processor with minimum control overhead and interconnect delay may achieve 100 MOPS/W at 1.25 micron geometry. In contrast a highly flexible, general processor, with a powerful high level language programming capability may only achieve 1 MOP. In general the more complex the processing task the more one must use high level programming techniques to facilitate system design and modification, and the greater the penalty in performance per chip. For this reason, and because of larger memory requirements, a complex multimode airborne radar processor with a throughput of about 600 MOPS may require more than 100 VHPIC chips whereas a homing head processor with a throughput of 100 MOPS may require only about 4 chips.

15.3 Defence applications of VHPIC

Defence has perhaps the most demanding requirements for very high performance, particularly in digital signal processing. Defence is heavily dependent on such functions as surveillance and intelligence, signature recognition, targeting and guidance, communications, electronic warfare, all of which involve intensive signal processing. Military systems have to operate in real time, generally with demanding constraints on the available power and space. There is almost no limit to the amount of information processing that could be beneficially used in military systems, if available at a suitable size, power consumption and cost. Indeed, today's problem in defence is often not how to acquire more information, but how to make use of the information already available without overwhelming the military command and control system. Making more use of information implies much greater signal processing, making full use of the available bandwidths, better methods of information reduction and development of powerful machine intelligence.

15.3.1 Classes of application

The 1982 VHPIC Programme Definition Working Group [5] divided military uses of VHPIC into 3 key classes: (i) modular processors with dedicated hardware, mainly for smart weapons; (ii) highly programmable high performance array processors; (iii) low power processors for portable equipment such as man-portable radios. These classes differentiated applications primarily in terms of the algorithms involved, the programmability required and the production scale and cost (Gray and Darby [6]). Underlying this analysis is the need to identify maximum commonality between applications in order to minimise the number of high cost VHPIC processors that have to be developed and updated, and the need to adopt a modular approach based on common operations which can be implemented as a chip or macrocell (section 6.2), so that a processor for a specific application can be tailored from pre-existing modules.

15.3.1.1 Weapons processors

Weapons are often manufactured in relatively large production runs, varying from tens of thousands for tactical missiles to millions for small mines. Production cost is therefore important, and since much of this cost lies in assembly and quality assurance, it is sensible to aim for a high level of integration, reliability and built-in test. It is also important that the processor should be proof against malfunctions that could cause the missile to home wrongly or the warhead to detonate spuriously due to nuclear radiation, EMP and other electromagnetic interference or due to fire and other severe stresses.

For some weapons where the signature analysis is relatively straightforward and predetermined, the processing can be handled by a signal processing microprocessor with a throughput of a few MOPS. However, highly intelligent homing heads may require throughputs of 100 MOPS or more, whilst retaining small size and reasonably low power consumption. This indicates the need for a hardware-efficient approach, with only limited software programmability. Ideally the processor should be sufficiently intelligent to be able to recognise targets generically so that it does not need to be programmed on specific target features which could be altered or camouflaged.

15.3.1.2 High performance, highly programmable processors

For many applications in ground installations and in ships, aircraft and other platforms, one requires a very high degree of programmability together with a throughput of 1000 MOPS or greater. This indicates using a general purpose processor, whose high design cost can be amortised over many different applications, and which, most importantly, provides a powerful high level programming environment to assist the development and adaptation of very complex processing routines. The processor can take many forms depending partly on the degree of data dependence involved and its effect on the connectivity (section 6.1). For example it may be a highly parallel processor array such as the ICL Distributed Array Processor (DAP), it may be a smaller array of more powerful processing elements such as the transputer, it may be a data-bus multiprocessor system such as Dipod or it may be a highly pipelined supercomputer. Many of these architectures have been under study at RSRE and elsewhere for several years to determine their relative merits and disadvantages (Roberts et al. [7]).

15.3.1.3 *Multimode airborne radar*

An example of an application requiring a highly programmable, very high throughput processor, is multimode airborne radar [6]. Future tactical radars must operate in up to 20 different modes including air-air target detection, acquisition and tracking and air-ground real beam, doppler beam sharpening, synthetic aperture radar, stationary and moving target detection, weapon delivery, automatic identity-friend-or-foe, terrain following, obstacle avoidance and precision navigation. To combine all these modes in a single multi-mode radar system, with the capability of changing modes in real time, demands very high data throughput and programmability.

15.3.1.4 *Low power portable systems – digital radio*

Digital processing provides the capability to make a highly programmable communications set. This gives flexibility to alter carrier frequency, modulation type, error control coding, encryption, network management, ECCM, and to cope with new EW threats during the equipment's life. Functions may include bandwidth compression, adaptive filtering for noise suppression, automatic speech recognition/generation and a schematic graphics communications capability. With such flexibility and power, the aim is to produce a radio suitable for all three Services, for man-pack and platform use, and inter-operable between different NATO groups, giving better performance plus cost saving from increased production.

15.3.2 *Avoiding obsolescence*

It may typically take 5 years for a military equipment to move from research into production, during which time IC technology may improve in performance by an order of magnitude, and the equipment specification may also have altered. In the past, whilst commercial equipments with shorter development times have used state-of-the-art ICs, military equipment often used IC technology which was already obsolescent by the time the equipment went into production. Today this is an unacceptable penalty, when processor performance is becoming such a key aspect and when one needs to achieve the highest performance/cost ratio. The failures of the past, however, show that this problem may be very hard to cure.

In many of the applications discussed in section 3.1 there is sufficient space and power at the system development phase for one to design an oversized processor in current technology with the intention of shrinking it with state of the art technology at the production stage. In some cases however this may be more problematic: if one is developing a total platform such as a new tank or aircraft all the processing cannot be oversized and overpowered even for trials. An alternative is to accept a poorer performance at trials with the intention of upgrading it at the production stage.

A further important point is that it will only be feasible to upgrade processors at the production stage if there are radical improvements in software and hardware design tools during the next 5 years that massively reduce the cost of processor redesign. This is a key issue for VHPIC and is discussed further in section 6.

15.3.3 Some options for an upgrade path to avoid obsolescence

15.3.3.1 Upgrading an existing machine

One possible route is to take a single well proven processor architecture and shrink it. This route has been used by RSRE and ICL in shrinking the first generation DAP into a compact version (MILDAP) with a 20 MOP performance suitable for prototype feasibility demonstration, with the possibility of taking the same machine to a VHPIC DAP with 1000 MOP performance. Processing developed on the early generation machine can then be carried directly onto the later machine. This is advantageous since upgrading software for a complex system can be quite as difficult as upgrading the hardware.

A similar approach is to use an array of identical powerful processor chips, such as the INMOS transputer, so that as system development proceeds into production, the availablity of more powerful chips can be exploited to reduce the number of chips needed in the final processor, whilst keeping the operation unchanged. The transputer is well conceived for this approach because it enables one to very easily repartition the overall computation to allow for changes in the number of individual processors and their connectivity. A transputer array is programmed in OCCAM as communicating processes and the decisions on the connectivity of the hardware array and the assignment of processes to processors is taken at the final stage, and is therefore easy to alter.

15.3.3.2 Migration of software into modular hardware

A second route is to adopt an architecture which facilitates migration of software into hardware, so that advanced IC technology can be used to upgrade machine performance by implementing key process functions in modular hardware as the system development proceeds. Dipod, developed by Logica and RSRE, is an example of such a machine, which uses an intermediate language, Fith, matched to the hardware architecture. Such a machine may assist rapid uptake of advances in areas such as image processing since these can be added as hardware modules without major disruption of the total system.

15.3.3.3 Emulation and virtual machines

One can use emulators to develop processing before the VHPIC processor itself is available: this however means that there may be no graceful fallback if the VHPIC machine fails to arrive on schedule. An alternative may be to emulate via a virtual machine or an intermediate compilation language, with more than one VHPIC processor being developed to implement the virtual machine. This approach is attracting interest in the Alvey architecture programme, but may not be able to achieve sufficient hardware efficiency to meet VHPIC requirements unless the virtual machine is very well suited for VLSI implementation. At present the cost of designing a really efficient VLSI machine is so high that one probably cannot afford multiple sources unless there are major civil applications.

15.3.3.4 Graceful fallback

If one aims to use state-of-the-art circuits, the problem of achieving graceful fallback is present whatever approach one adopts. As often as not, a new IC process does not reach production on time, particularly if it incorporates some major advance such as an additional level of interconnect. One possible way to protect against these slippages is to structure the design hierarchy to give maximum process independence at each design level (section 6.2). This could facilitate changing process if this is necessary to meet project timescales.

15.4 VHPIC as a cost reducing technology

In addition to its higher performance, VHPIC is also important for its potential to reduce military equipment cost, just as in the civil sphere improvements in LSI/VLSI technology have led to remarkable cost reductions, exemplified by the personal computer. There are many ways in which VHPIC can reduce Defence costs. The following are some examples. Production costs are substantially reduced if the whole processor can be integrated on a single circuit board or a single chip, eliminating the cost and unreliability associated with circuit board assembly and interconnection. Self-testing and self-repair may reduce the high cost of military quality assurance and maintenance. Electronic replacement of mechanical parts offers major savings: this includes electronically steered sensors to replace mechanical tracking. Incorporation of processing in sensors may permit pyroelectric and other room temperature infra-red detectors to replace cryogenic detectors, and may radically reduce other sensor costs. Improvements in weapon accuracy can permit smaller warheads making the weapon lighter and cheaper and allowing more to be carried. Improvements in autonomous homing can reduce the sophistication and cost of weapon platforms. Replacement of manned vehicles and aircraft by intelligent unmanned vehicles may radically reduce equipment costs by permitting major improvements in design, mobility and reduction of armour. Increased commonality with civil information technology systems will also help to reduce development and production costs of military systems, and it is pertinent that many civil systems achieve much better reliability than their military counterparts, despite the very much higher cost and exhaustive development of the military systems.

15.5 Civil applications of VHPIC

The linking of the MOD VHPIC technology programme into Alvey reflects the strong commonality between civil and military requirements. There are of course a number of areas in which military equipment does have rather special requirements. Perhaps the most unequivocal of these is environmental, the need for -55 to $+125$ degrees C temperature range and an increasingly mandatory requirement for nuclear hardness. But the distinction between military and commercial requirements can easily be exaggerated, and is probably dwindling.

15.5.1 *Digital signal processing*
There are many commercial products that require high performance signal processing, and this will increase in the future. These application areas include automobile electronics, interactive graphics, computer aided manufacture, process equipment and industrial plant control, instrumentation, medical prosthetic, orthotic and biofeedback devices, computer aided tomography and NMR imaging, civil radar and air traffic control, satellites and aerospace [8].

15.5.2 *Pattern recognition*
Image understanding is of key importance for computer vision and robotics [8]. Military uses perhaps provide the most demanding research lead because of the high background clutter characteristic of military images. But it may be the civil applications that provide the major technology development and cost reductions, which can in turn be exploited by Defence. Automatic speech synthesis and recognition is another area where military and civil goals are similar.

15.5.3 *Digital communications*
In digital communications there is a strong commonality with commercial applications, and it may perhaps prove significant for Defence interests that one of the demonstrator projects being funded under the Alvey programme is a mobile information system for civil vehicles. Many civil and military systems have a common need for encryption and secure communications. Through developments in information technology in the commercial field, there will be widespread use of computer-based information, which will involve quite sophisticated mobile communications networks. Likewise military communications will be affected by the need to transmit larger amounts of computer based information round the battlefield.

15.6 VHPIC design

VHPIC technology at a feature size of around 1.25 micron will make possible chip complexities of more than a hundred thousand gates. At present the cost of designing chips of this complexity is very high, typically £100 per gate for custom design. The design time is also several years. The feasibility of VHPIC circuits of this complexity therefore depends critically on improvements in computer aided design and particularly on semicustom and cell-based design through which design time and cost can hopefully be improved by at least an order of magnitude.

At the system design level VLSI and VHPIC creates the capability to produce systems so complex that the management of that complexity arguably dwarfs all other issues.

15.6.1 Design optimisation

In the past hardware minimisation was generally the dominant factor in electronic design. Today, design cost and connectivity are becoming the two most important considerations. But there are many factors which need to be taken into account including the trade-offs between high throughput, programmability and power consumption, choice of algorithm, choice of parallelism, regularity of data and control flow and of the hardware structure, timing, hardware and software validation and compliance with portability standards.

15.6.1.1 Semicustom versus standard processor

A basic decision confronting the VHPIC system designer is to what extent to use a standard processor and to what extent to use semicustom design based on PLA, gate array and macrocell. This is a choice between software design and hardware design. If a high throughput per chip is required the choices may be between microcode on the software side or cell-based custom design on the hardware side, or often both. But if the problem is very complex a software route based on a high level language may be the only sensible approach at the present time. This means some sacrifice in hardware performance, but careful attention to algorithms, judicious choice of machine and language, and implementation of key functions in hardware can all assist in minimising any performance loss. For example the Intel iAPX432 chip set implements much of its operating system kernel in hardware in order to run ADA efficiently. Further improvements can be obtained by a limited degree of specialisation of the processor architecture for signal processing operations, such as the TI TMS 320 and the STL CRISP. VHPIC technology will allow such processors to be produced with throughputs of around 8 MOPS/chip.

15.6.1.2 Choice of parallelism

It is common adage that VLSI equals parallelism. However this is perhaps oversimplistic. VLSI implies complexity, and complexity implies the need to handle tasks in parallel. But complexity also implies the need to limit communication. Otherwise chip area will become consumed by metal interconnect and interconnect capacitance will erode circuit speed, even assuming that advances in multilevel metal technology allow several further levels of on-chip interconnect. There is therefore a conflict: parallelism tends to increase interconnect, because one is handling a smaller amount of the computation in a single physical location before passing the data to another location. One solution is to build powerful processing nodes incorporating extensive memory to minimise the requirements for node-node communication. This is the approach adopted in transputer arrays, for example. An alternative solution is to make very heavy use of pipelining, either 1-dimensional pipelining as in vector supercomputers, 2-dimensional pipelining as in a systolic array processor. Pipelining can, allow very regular nearest neighbour connectivity, and this allows one to achieve the high communication intensity required for parallel processing whilst keeping the interconnect overhead acceptable.

15.6.1.3 *Impact of VHPIC on the systems engineer*

One of the benefits of using a new architecture such as the DAP is that the system engineer is forced to think very carefully about his application. All too often engineers programming on a conventional machine in a high level language do not think sufficiently fundamentally, and are in fact using the machine inefficiently, without realising this, as for example when programming a vector supercomputer in a sequential high level language.

But, perhaps the most important benefit of VHPIC and VLSI on system design will be to enable system engineers to evaluate new concepts, which currently are stalled because the cost and effort of implementing them is so high. Two crucial advances are needed. Better CAD must radically reduce the design cost and timescale. Secondly, the systems engineer must be able to manipulate designs in silicon without needing to become an expert VLSI circuit engineer. He needs to be able to build new systems from high level building blocks, where the building blocks themselves encapsulate the bottom-up design optimisation by the VLSI engineer in a transparent way. The system engineer needs powerful tools so that the design complexity does not overwhelm him. This is one of the great promises offered by semicustom design, particularly macrocell.

15.6.2 *Macrocell design*

Macrocell design is a mixture of hardware and software design. Small cells will be put together to produce structures that are entirely hardware defined. But as cells become larger they will inevitably be more specialised unless they have at least microcode programmability. Libraries of such cells will evolve, together with libraries of associated microcode and higher level software. Macrocells will also be 'parametrised' so that their hardware performance can be tailored to new applications. Automatic tools will exist to position, stretch, compact and

Very high performance integrated circuits 129

interface macrocells to make powerful ICs. However these techniques are still in their infancy.

As chip complexity increases macrocells will probably become the optimum IC design technique, provided that adequate tools exist to allow macrocells to be genuinely portable and reusable. The issue of macrocell portability is discussed in the next section.

15.6.2.1 *Exploitation and exchangeability of macrocells*

At the highest level, processors can be described behaviourally, independent even of their functional implementation in chips and macrocells. At the lowest level they can be described as layouts. The highest level is very hardware independent. In contrast, at the mask level the macrocell design can only be reused to fabricate an identical processor on a process with comparable design rules. Between the highest and lowest levels of description there are intermediate levels, chip and macrocell function, register transfer, gate, transistor and sticks, which provide intermediate degrees of reusability. For example, the macrocell function level might provide an international standard macrocell library, the macrocell register transfer and gate level designs might be reused on a different process; a transistor level design might be reused on a similar process, but having substantially different design rules and different geometry; a sticks level design might be reused on a similar process but having somewhat different design rules. Macrocells are a key focus of the Alvey Programme, and the strong Alvey emphasis on industrial collaboration will be particularly valuable in the macrocell area in defining the right standards and in achieving wide exploitation.

15.6.3 *Gate arrays*

Gate arrays are now becoming available with very powerful design tools, providing fully automated placement and routing, and in some cases simulation and test pattern generation, with complexities up to 20,000 gates. This goes a long way to meeting the goal of a user friendly design capability, which can be used by system engineers unskilled in VLSI layout. In addition greater computer control of process lines will enable gate array vendors to handle much smaller gate array production runs profitably. Gate arrays do not give the ultimate in performance, however, and they can prove a very expensive design route if near state of the art performance is required. But provided the performance is not too demanding greater use of gate arrays in military and civil systems would offer major benefits in compactness, reliability and lower cost. Gate array design is less flexible than macrocell. It is best suited for implementing logic. As chip complexity increases an intermediate semicustom approach, mid-way between gate array and macrocell may also emerge consisting of gate array along with a limited range of macrocell function blocks, such as memory, or signal processors.

15.6.4. *Testability and self-testing*

Not least in significance is the problem of VHPIC testing and design for testability, including self-test and fault tolerance. This was singled out as one of the most important areas for VHPIC CAD research in the joint RSRE-industry study on VHPIC CAD in early 1983, along with CAD interface standardisation, system level description languages and macrocell libraries and manipulation software.

15.6.5 CAD interface standards

The problem of collaboration and exchangeablility of macrocells discussed in section 6.2 is one example of the much wider problem of achieving interface standards in CAD. The need for collaboration in CAD is very apparent: the UK CAD community is too small to be able to support the production of many incompatible design systems. For this reason in 1983 a CAD Interface Standardisation Project (CISP) was funded as the first stage of the VHPIC CAD programme (now subsumed into Alvey). CISP was carried out by 6 teams drawn from 15 organisations covering methods of collaboration, software engineering for CAD, disc and instore representation, design representation and interchange, and graphics representation, and provided an important base from which to progress to greater unity and technology sharing in CAD.

15.7 Conclusion

VHPIC, information technology and VLSI technology are of key importance to military and civil equipment and systems. But it is a highly demanding area, in investment, expertise and new thinking. It will be very beneficial to exploit commonality between applications, both civil and military. On the military side cost reduction through use of VHPIC may be as important as increased performance. Improved design tools to give an order of magnitude reduction in design costs will be crucial. Effective continuing collaboration between companies to share the research task and to exchange technology, design tools and VHPIC circuits, will be very important to success.

15.8 References

1. The VHSIC Program, Office of the Undersecretary of Defense for Research and Engineering, Washington DC, April 1984.
2. VHSIC Bulletin, OUDRE, 1984/4, published by Palisades Institute, 201 Varick St., NY 10014, USA.
3. Alvey VLSI and CAD Strategy, DTI, Millbank, London SW1 4QU, 1984.
4. MEARS, A. L. VHPIC, Proceedings 1982 Military Microwave Conference, London, 20-22 October 1982, pp. 592–597.
5. VHPIC: A Proposed MOD Programme, RSRE, June 1982 (Unpublished MOD Report).
6. GRAY, K. W., and DARBY, B. J., Future Requirements for VLSI in the Military Field, Systems on Silicon, pp. 85–88, Peregrinus, 1984.

7. ROBERTS, J. B. G., SIMPSON, P., MERRIFIELD, B. C., and CROSS, J. F., Signal Processing Applications of a Distributed Array Processor, to be published Proc IEE Part F, Special Issue on DSP, October 1984.
8. VLSI Electronics Volumes 2, 3, 4 and 5, Editor Einspruch N. G, Academic Press, 1981/1982, ISBN 0-12-234102, 3, 4, 5.

Copyright © Controller HMSO, London 1985

Chapter 16
Very high speed integrated circuits (VHSIC) technology for digital signal processing applications

Carl W. Monk, Jr.
VHSIC/Electron Devices

16.1 Introduction

The Very High Speed Integrated Circuits (VHSIC) Program of the United States Department of Defense will have very significant impact on present and future digital signal processing capabilities. Just as English forces used the longbow at the battle of Crecy in 1346 to multiply their effectiveness, so must modern Western military forces multiply their effectivness through the use of advanced technology, particularly electronics technology. The VHSIC Program will preserve and extend the Western lead in deployed electronics technology, providing improvements in signal processing capability for surveillance, communications, target classification and several other appliction areas where technology can overcome numerical advantages of potential adversaries.

16.2 Overview of the VHSIC program

16.2.1 *The need for VHSIC*

Through the decade of the 1970's, the United States Department of Defense (DoD) followed a deliberate policy of not providing significant funding for programs to advance the state-of-the-art of integrated circuits (IC's). The rationale for that policy was based upon the fact that the IC industry was continuing to make advances in support of its market requirements, and that the DoD benefitted from those advances.

Over that period, a different situation has evolved, necessitating a major reversal of this past hands-off policy with regard to IC's. First, the DoD began to find itself with less and less "clout" in the IC marketplace. With the advent of the microprocessor and the emergence of myriads of commercial applications for digital integrated circuits, the DoD share of the total IC market in the

VHSIC Technology for digital signal processing applications   133

United States has slipped to less than 10 per cent. Increasingly, the U.S. semiconductor industry has been reluctant to qualify their IC's to military specifications. Furthermore, concepts such as built-in-test (BIT) and fault tolerance, which are extremely important to many military applications, have not been developed and applied. To a large extent, the signal processing needs of the DoD have not received adequate attention.

Secondly, the transition of IC technology from large scale integration (LSI) to very large scale integration (VLSI), with the capability to fabricate very large numbers of equivalent gates on a single silicon chip, is introducing an era of *integrated systems*. In order to exploit this new era to the maximum advantage, we must develop new ways of thinking about silicon chips. Specifically, the designer of chips can no longer perform just the classical functions of circuit configuration, but must know and apply the principles of operation and design of the end-item system application.

Finally, as gate complexity levels have increased, military IC's have more frequently been custom designed. This trend, coupled with the integration of many functions onto fewer and fewer IC's in a system, has resulted in VLSI becoming very expensive to the DoD.

During this same time, coincidental with the emergence of the "all-volunteer Army" in the United States, it became increasingly important for the United States to evolve a way of multiplying the effectivness of its military forces through the use of technology in general and electronics in particular. Nowhere was this need more apparent than in the areas of target sensing, classification and tracking, and in communications. It was apparent that underlying each of these areas was the need for advanced signal processing technology. Out of this need was born the VHSIC Program.

16.2.2 Objectives of the program

The Very High Speed Integrated Circuits Program was initiated in 1979 to end the hands-off policy toward integrated circuits. Heading the list of VHSIC Program objectives is the acquisition of advanced signal processing capabilities for military systems through tailored application of advanced IC technology. This is followed closely by objectives to provide IC's qualified to military specifications, to reduce the time delay required to insert newly developed technology, and to reduce life cycle costs. Specific goals were established in these areas:

- Increase in functional throughout (one to two orders of magnitude).
- Ease of technology insertion into weapon systems.
- High tolerance to radiation environments.
- Improved reliability and testability.
- Minimum life cycle costs.

To meet these goals during the 1980's and early 1990's, the DoD decided to focus on digital silicon IC technology. Because of the wide variety of weapon system needs, the tri-Service VHSIC Program elected to include as many major technologies as possible, including both bipolar and CMOS. Moreover, it was determined that several major IC design styles would be needed, so provisions were made for custom as well as semi-custom, and programmable as well as non-programmable, design approaches.

To help ensure that the VHSIC technology could, and indeed would, be inserted into DoD weapon systems in a timely way, each chip development effort was required to include a demonstration of applicability to a major weapon system through development of a brassboard (or prototype) subsystem. This early VHSIC emphasis on "built-in" technology insertion has paid off handsomely in terms of providing direction for VHSIC technology evolution.

The VHSIC brassboards cover a wide range of signal processing applications, including radar, sonar, electronic warfare and communications.

16.2.3 VHSIC program structure and status

The VHSIC Program has been structured into four phases, numbered zero through three. Each phase of VHSIC has been characterized by careful planning and well-defined, specific goals. VHSIC Phase 0, conducted in 1980, was essentially a government/industry planning exercise to define VHSIC Phase 1. VHSIC Phase 1, which began in 1981 and is drawing to a conclusion now, encompasses the design and fabrication of sets of IC's with minimum feature sizes of 1.25 micrometers, the development of a pilot production capability for those devices, and a demonstration of the applicability of the devices through a signal processing application brassboard. Six contractors were selected to participate in Phase 1. Fig. 16.1 and 16.2 show the direction of the programs of each of the six contractors, and the chip sets being developed.

VHSIC Phase 1 is nearing completion. The program is on target and VHSIC 1.25 micrometer chips and signal processing brassboards using these chips are even now being demonstrated by the VHSIC contractors to their tri-Service sponsors. Nearly 2,000 fully functional prototypes of 11 of the 29 Phase 1 chip types have been fabricated and tested, and the fabrication of most of the remaining types has begun. Each Phase 1 contractor has produced at least one fully functional device from its chip set.

In parallel with the VHSIC Phase 0 and Phase 1 efforts has been the VHSIC Phase 3 program. The aim of this portion of the VHSIC Program has been to augment and technically underpin the Phase 1 efforts with new and innovative approaches from the industry and university research communities. During the first four years of the VHSIC Program, the Phase 3 effort has been feeding up-to-date ideas and approaches into the mainstream VHSIC effort in the areas of starting materials, lithography, packaging, testing, built-in test and fault tolerance, and algorithms.

VHSIC PHASE I CONTRACTORS

CONTRACTOR (SUBCONTRACTOR)	TECHNOLOGY	BRASSBOARD	DESIGN APPROACH	SPECIAL FEATURES
Honeywell	BIPOLAR — CML, ISL	ELECTRO-OPTIC SIGNAL PROCESSOR	CUSTOM CHIPS BASED ON MACROCELL LIBRARY	RADIATION HARD MULTI-SITE CAD NETWORK
Hughes (Perkin-Elmer)	CMOS/SOS	AJ COMMUNICATIONS	CUSTOM RECONFIGURABLE CHIPS	RADIATION HARD HIGHLY SPECIALIZED CHIPS E-BEAM MACHINE DEVELOPMENT
IBM	NMOS	ACOUSTIC SIGNAL PROCESSOR	MASTER IMAGE WITH MACROCELL/MICROCELL LIBRARY	SOFTWARE STRENGTH HIGH DENSITY CHIP DESIGN
TI	BIPOLAR — STL NMOS	MULTIMODE FIRE-AND-FORGET MISSILE	PROGRAMMABLE CHIP SET	OPERATIONAL FABRICATION FACILITY DESIGN UTILITY SYSTEM
TRW (Sperry-Univac) (Motorola)	BIPOLAR — 3D T²L CMOS	ELECTRONIC WARFARE SIGNAL PROCESSOR	STANDARD CHIP SET	INNOVATIVE MEMORY VERSATILE CHIP SET
Westinghouse (National) (Control Data) (Harris) (CHU)	CMOS/BULK	ADVANCED TACTICAL RADAR PROCESSOR	STANDARD CHIP SET	MULTI-CHIP FUNCTIONAL MODULES ARCHITECTURE

Fig. 16.1 *VHSIC phase 1 contractors*

VHSIC CHIP SETS

HONEYWELL
PARALLEL PIPELINE PROCESSOR
SEQUENCER
ARITHMETIC UNIT

TI
DATA PROCESSOR UNIT (1750A)
VECTOR ARITHMETIC/LOGIC UNIT
STATIC RAM
MULTIPATH SWITCH
VECTOR ADDRESS GENERATOR
ARRAY CONTROLLER/SEQUENCER
DEVICE INTERFACE UNIT
GENERAL BUFFER UNIT

IBM
COMPLEX MULTIPLIER
ACCUMULATOR

HUGHES
DIGITAL CORRELATOR
ALGEBRAIC ENCODER/DECODER
SIGNAL TRACKING SUBSYSTEM

TRW
CONTENT ADDRESSABLE MEMORY
WINDOW ADDRESSABLE MEMORY
REGISTER ARITHMETIC LOGIC UNIT
ADDRESS GENERATOR
MATRIX SWITCH
16-BIT MULTIPLIER/ACCUMULATOR
MICROCONTROLLER
FOUR-PORT MEMORY

WESTINGHOUSE
EXTENDED ARITHMETIC UNIT
MULTIPLIER
PIPELINE ARITHMETIC UNIT
EXTENDED ARITHMETIC UNIT
CONTROLLER
10K GATE ARRAY
STATIC RAM

Fig. 16.2 *VHSIC chip sets*

VHSIC Technology for digital signal processing applications

16.2.4 Current efforts

16.2.4.1 VHSIC Phase 2

Now, a new phase of VHSIC is beginning. VHSIC Phase 2 aims toward the realization of submicrometer technology, in a fashion similar to the Phase 1 effort. Proposals have been evaluated and contract awards are imminent for this step in the development of VHSIC. Winning contractors will design and fabricate chip sets with 0.5 micrometer feature sizes, develop pilot production capabilities, and demonstrate their chip sets in signal processing brassboards.

In parallel with the submicrometer technology development under VHSIC Phase 2 will be three major thrusts that have grown out of the Phase 1 progress: a yield enhancement effort, a technology insertion program, and the development of an Integrated Design Automation System (IDAS).

16.2.4.2 Yield enhancement

Each of the Phase 1 contractors has been tasked with an additional 30-month effort to increase the yield being obtained in fabrication of their Phase 1 devices. The objective is to identify the eliminate yield detractors and achieve an order of magnitude increase in fabrication yield, and thus a concomitant order of magnitude reduction in the cost of the devices to defense contractors desiring to incorporate them into new weapon system designs.

16.2.4.3 Technology insertion

As with any new technology, the use of VHSIC in current systems developments will present system program managers with risks of cost escalation and schedule slips. In order to minimze these risks, the DoD VHSIC Program Office is providing seed money for joint funding with system managers of ventures which will enable them to pursue VHSIC technology in parallel with more conventional approaches. Approximately thirty systems in current development in the three U.S. military services have been selected to receive these VHSIC technology insertion funds. The following list of technology insertion programs initiated in 1983 includes systems that cover almost all type weapon platforms and most major military missions, including surveillance, communications, electronic warfare, and guidance and control.
Army:
- TOW-2 automatic target tracker
- M-1 tank, signal processor for fire control
- Airborne signal processor for LHX helicopter
- Joint Service Seeker application to Hellfire
- Threat signal processor for Patriot's HIMADS radar

Navy:
- Acoustic signal conditioner for AN/UYS-1 acoustic signal processor
- Programmable signal processor for AN/APG-65 radar
- AN/AYK-14 standard airborne computer

138 VHSIC Technology for digital signal processing applications

- MK-50 Advanced Light Weight Torpedo
- Floating-point arithmetic processor for Enhanced Modular Signal Processor

Air Force:
- AN/ALQ-131 electronic warfare pod
- Modular Avionics Design, advanced standard modules
- Launch and Leave Guided Bomb
- 1750A general purpose computer
- Common signal processor

16.2.4.4 *IDAS*

An Integrated Design Automation System (IDAS) is necessary to reduce the time it takes to develop or modify the design of complex IC's. The VHSIC IDAS effort is in its first phase, which has the limited objective of capturing the automation technology employed to date in the VHSIC Program, and building upon it to produce more advanced computer aided design tools. The long term objective of the IDAS effort is to develop a fully integrated hierarchical design system covering abstraction levels from system partitioning through logic design and mask generation.

In summary of this overview of the program, the first phase is nearing completion and the yield enhancement, technology insertion, IDAS and Phase 2 submicrometer technology development thrusts are all underway. Fully functional chips have been produced, and signal processing system brassboards are now being demonstrated. The military Services have vigorously undertaken the task of inserting this new technology into a wide variety of signal processing systems.

16.3 Signal processing applications of VHSIC

16.3.1 *Advantages of digital signal processing*

How does VHSIC fit into the overall context of the digital signal processing evolution, and what will be the payoffs of employing VHSIC? To examine the place of VHSIC technology in signal processing, consider a generic model of a signal processor: its components are a sensor, a conversion module, a processing module, a target detecting/classification module, and a display module. The portion of the system which can be realized with digital technology is increasing with time. More and more analog signal processing functions are being implemented in digital circuitry, and the analog/digital boundary is moving toward the sensor front-end of the generic model. VHSIC can accelerate this trend, with a dramatic impact in terms of improvements in system performance, maintainability and reliability.

In particular, digital (versus analog) signal processing technology offers advantages of flexibility, wider dynamic range, reproducibility, freedom from adjustment, and self-testing fault-tolerance. The graph of Fig. 16.3 gives an idea

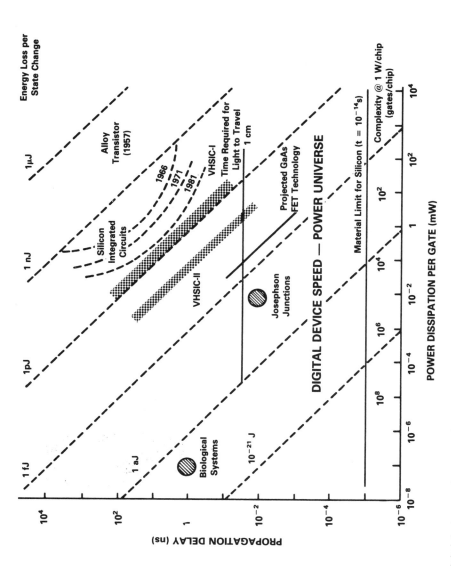

Fig. 16.3 Digital device speed-power universe

of how VHSIC compares with other digital technologies. At the beginning of the digital integrated circuit era in the mid-1960's, minimum IC feature sizes were on the order of 20 micrometers, resulting in a speed-power product of perhaps 100 picojoules. (The speed-power product used here is the product of power dissipation per gate and gate propagation delay.) By the early 1970's feature sizes had been reduced to about 10 micrometers, and by the early 1980's, to 2-3 micrometers, with speed-power products of about 10 picojoules.

16.3.2 Impact of VHSIC

Fig. 16.3 shows how VHSIC technology (Phase 1 – 1.25 micrometer, and Phase 2 – 0.5 micrometer) compares with other technologies. Phase 2 will provide two orders of magnitude speed-power improvement over the 2–3 micrometer VLSI, although the technology will still fall short of projected gallium-arsenide field effect transistor technology. The impact at the signal processing system level will be quite dramatic in terms of the kinds of complex functions that can be performed.

Major life cycle savings will accrue through the use of VHSIC. Significant improvements in system size, weight, and power consumption will be realized, and at the same time reliability will increase due to reduced parts counts and fewer interconnections in electronics systems. An overall payoff may be the achievement of truly "smart" weapons and systems. Such future applications may include:

- Ground-based radar that can distinguish between friendly and enemy aircraft by performing a fine-grained intra-pulse analysis of radar returns in real time.
- Airborne synthetic aperture radars capable of providing guidance for radar-guided missiles during own-aircraft evasive manoeuvres.
- Fire-and-forget missile guidance subsystems.
- A small and low cost manpack providing precise positioning and allowing preset messages to be selected and transmitted in a secure format in a netted system.

Immediate improvements can be made with Phase 1 VHSIC devices, such as for these common signal processing systems: radar, passive sonar, and communications. We will begin with replacing older technology with VHSIC. Examples are analog charge-coupled devices and surface acoustic waves in radar systems. In addition, radar systems with digital signal processing will be able to spot targets with much smaller cross sections in higher clutter environments. Target location will be more accurate because of improved beamforming accuracy. Many more targets can be tracked simultaneously with the increased processing power of VHSIC. Radar systems can be made more compact for platforms where size, weight and power are critical parameters.

In the passive sonar signal processing arena, many more channels of acoustic data could be processed. Significantly improved and lower cost adaptive beamforming, capable of detecting targets in noisier environments, will be possible

VHSIC Technology for digital signal processing applications 141

with VHSIC technology. Because more effective (but more complicated) algorithms can be processed in real-time with VHSIC, greatly improved target detection and classification will be possible.

In new communications systems realized using VHSIC technology, more data and voice channels can be processed, more reliably and with much more resistance to jamming.

All in all, VHSIC offers much to the signal processing community. It is up to that community to make maximum use of VHSIC.

16.4 Design automation

To aid the signal processing community in the utilization of VHSIC technology, the VHSIC Program has created numerous computer aided design (CAD) tools. Now these CAD tools are being pulled together and made available to Defense Department contractors as a means of dealing readily with VHSIC technology.

The IDAS effort has four major aspects. First, a VHSIC Hardware Description Language (VHDL) is being defined to provide a common high-level descriptive medium for all VHSIC chips. Secondly, a three-part program, not unlike the present VHSIC technology development effort itself, has been defined. The goals of these parts are distinct, but related. Under IDAS-1, existing CAD tools will be identified and evaluated. Under IDAS-2, these tools will be integrated around the VHDL and a common data base. In parallel with IDAS-1 and IDAS-2, development of the next generation of fully automated and integrated VHSIC CAD tools will begin under IDAS-3.

The IDAS-1 and -2 effects will focus on collecting most of the types of tools presently found in the VHSIC program. The aim of IDAS-3 is quite different and extremely important. IDAS-3 is directed toward more advanced design automation. IDAS-3 will be characterized by advanced user interfaces, a fully integrated data base, and extensive use of expert system technology (or artificial intelligence).

As 1984 draws to a close, VHSIC Phase 1 is also drawing to a close with some very exciting successes behind. Beyond this juncture is the submicrometer VHSIC Phase 2 effort and an exciting VHSIC Phase 1 technology insertion program which includes the development of design automation tools which will allow DoD signal processing system designers to fully capitalize on VHSIC technology.

16.5 Future digital signal processing advances

From where will the next generation of improvements in signal processing arise? They will come out of carefully considered, interdisciplinary research. This research will include advances in technology such as we are seeing with VHSIC.

A great deal remains to be gained through improved algorithms and their efficient mapping onto sound architectures using advanced design automation tools (perhaps knowledge-based expert systems). But the greatest gains for the signal processing community will result from the successful integration of technology (both circuit and design) with algorithms and architectures.

As gunpowder came after the English longbow in history, so the integration of these factors will follow VHSIC and will result in more effective electronic systems for defense.

In summary, the VHSIC Program was conceived as a way to leverage electronics technology for defense. Signal processing was chosen as the specific vehicle to demonstrate this, with the result that a wide variety of signal processing requirements can be met with VHSIC technology. The flexibility of VHSIC technology is being demonstrated in signal processing systems built for that purpose.

Chapter 17
Digital filters

A. C. Davies
Centre for Information Engineering, The City University, England

17.1 Introduction

Several technologies are making signal-processing in a discrete-time format advantageous. Some involve analogue samples (charge-coupled devices, surface-acoustic-wave filters, switched-capacitor filters), while in others, such samples are quantized, processed in digital form, and subsequently re-converted to analogue form. Collectively, all these approaches may be referred to as *digital filters*, and if finite-wordlength effects are ignored, they are *linear discrete-time systems*.

Their structure can be described by discrete-time state-equations:

$$x_{k+1} = Ax_k + Bu_k$$
$$y_k = Cx_k + Du_k$$

u, x, y are column vectors which denote respectively inputs, state-variables and outputs. A, B, C, D are matrices which characterise the interconnections and scaling operations and suffix k denotes the kth sample-time.

These difference equations are in one-to-one correspondence with the system structure when this is built from adders, scalors and delay elements, and may be regarded either as the specification of a "wiring diagram" or as a computer algorithm which implements the system.

An nth order system has n delays, and the state-variables are the outputs of these delays. The dimension of x and of A is thus equal to the system order. Only single-input single-output systems will be considered, for which u and y are scalars.

Difference equations may be transformed to algebraic equations by the z-transform: denoting transforms of input and output sequences respectively by $U(z)$ and $Y(z)$, the transfer function $H(z)$ may be expressed as

$$H(z) = Y(z)/U(z) = C(zI - A)^{-1}B + D$$

The denominator polynomial is $|zI - A|$, so that the poles are eigenvalues of A. In uncontrollable or unobservable systems, eigenvalues which are not poles

144 Digital filters

can occur – corresponding to cancellation of numerator and denominator factors of $H(z)$ – but this is unlikely in the context of digital filters.

$H(z)$ is a rational function of z, normally written in terms of powers of z^{-1} since z^{-1} represents delay by one sample-time and is a physically realizable element.

For algebraic simplicity it is convenient to normalize the sampling frequency to unity, so z^{-1} denotes delay by one second, equivalent to $\exp(-s)$.

For $s = j\omega$, $\exp(-s) = \cos\omega - j\sin\omega$, so that for all discrete-time systems, $H(j\omega)$ is periodic in ω and the normalized frequency response in the range $-0.5 < f < +0.5$ is repeated along the whole real-frequency axis.

The squared magnitude is given by

$$|H(j\omega)|^2 = H(z) \cdot H(1/z)|_c$$

where c denotes evaluation on the unit circle, e.g. for $z = \exp(j\omega)$. $|H(j\omega)|^2$ is thus an even periodic function.

17.2 Structures

For a given transfer function, $H(z)$, there are many alternative structures; formally they may be related by a linear transformation of the state-variables. Thus, if $x = Tg$, where T is a non-singular matrix, g is a new set of state-variables, corresponding to new state-equations:

$$g_{k+1} = (T^{-1}AT)g_k + (T^{-1}B)u_k$$

$$y_k = (CT)g_k + Du_k$$

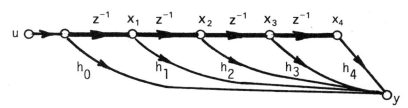

Fig. 17.1 *Non-recursive filter*

The structure is different but $H(z)$ is unchanged. Particular structures (direct form, parallel form, cascade form) are related by specific transformation matrices.

A non-recursive filter (transversal filter) has no feedback paths and is therefore necessarily stable, with a finite number of samples in its impulse-response (Fig. 17.1); its transfer function is a polynomial in z^{-1}, so all poles are at $z = 0$. This corresponds exactly to discrete convolution:

$$y_k = \sum_{i=0}^{n} h_i u_{k-i}$$

A recursive filter, having one or more feedback paths, can have poles outside the unit-circle of the z-plane, and can therefore be unstable for certain coefficient values. Of various possible structures (Fig. 17.2), a cascade of second-order sections is the most commonly used, since it generally has the lowest sensitivity to finite wordlength effects (limited precision of coefficients or rounding in

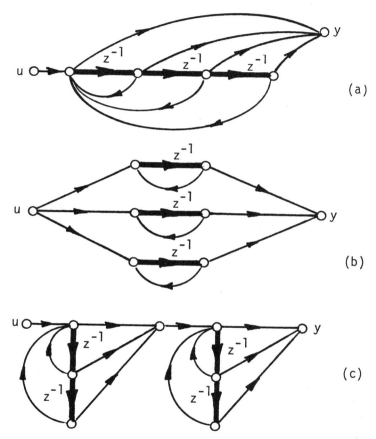

Fig. 17.2 *Some recursive filter structures: (a) direct (b) parallel (c) cascade of two second-order sections.*

arithmetic operations). Second-order rather than first-order sections are used so that complex-conjugate poles (or zeros) can be paired, avoiding complex coefficients and a need for complex arithmetic which would otherwise occur. The impulse response of a stable recursive filter typically has an exponentially-decaying envelope, giving an infinite number of samples, although recursive structures (such as the 'frequency-sampling' filter) with a finite impulse response are possible.

146   Digital filters

17.2.1 *Wave digital filters*

The magnitude vs. frequency response of passive lossless filters with equal resistive terminations has an exceptionally low sensitivity to component tolerances. This arises in part because at every attenuation zero in the passband, the first-order sensitivity of the insertion-loss is zero. Wave digital filters are special structures which attempt to duplicate this property. A direct simulation of lumped-element lossless filters leads to structures containing delay-free loops, which cannot be implemented. However, either by using scattering (wave) variables or by simulating unit-element (transmission-line) filters, digital filter structures can be obtained which imitate the low-sensitivity properties of passive lossless filters. Many such wave digital filters have a large number of multipliers and are impractical. However, some are capable of being multiplexed, offering economic implementation.

17.3 Design of infinite impulse response filters to meet piecewise-constraint magnitude-versus-frequency specifications

Since the specification of a digital filter must be periodic in frequency (as shown in Section 17.1), it is convenient for design purposes to transform the specification to a non-periodic one, in such a way that the 0 to 0.5 range of discrete-time systems is stretched to the 0 to ∞ range of continuous-time systems. This enables continuous-time filter design methods and design tables to be used directly for digital filter design.

A suitable conformal transformation is bilinear:

$$w = u + jv = c\frac{z-1}{z+1}$$

where c is an arbitrary constant. This maps the unit-circle of the z-plane into the whole of the jv axis. The periodic specification in ω (or f) becomes a non-periodic specification in v.

The simplest choice for c is unity, which makes one radian/sec on the v-axis equal to one-quarter of the sampling frequency (Fig. 17.3).

$$w = u + jv = \frac{z-1}{z+1} = \frac{1-z^{-1}}{1+z^{-1}} = \frac{1-\exp(-s)}{1+\exp(-s)} = \tanh\frac{s}{2}$$

The transformation "warps" the real-frequency axis, since $jv = \tanh(j\omega/2) = j\tan \pi f$.

Because of this warping, phase linearity is not preserved in the transformation, and the bilinear transform cannot easily be used for design of filters to meet phase or group-delay specifications.

The usual frequency-transformations (low-pass to high-pass or band-pass, etc.) may be carried out either in the z or w planes. The latter is more convenient, since it corresponds exactly to the procedure used in continuous-time filter design.

A filter may be designed in the traditional manner in the w-plane, giving a transfer function $H(w)$, from which $H(z)$ is obtained via the bilinear transformation.

For example a second-order section with w-plane zeros at infinity has the form

$$H(w) = \frac{k\omega_n^2}{w^2 + 2\zeta\omega_n w + \omega_n^2}$$

which transforms to

$$H(z) = \frac{k}{A}\left\{\frac{1 + 2z^{-1} + z^{-2}}{1 + \frac{C}{A}z^{-1} + \frac{B}{A}z^{-2}}\right\} \tag{17.1}$$

where

$$A = 1 + \frac{2\zeta}{\omega_n} + \frac{1}{\omega_n^2}, \quad B = 1 - \frac{2\zeta}{\omega_n} + \frac{1}{\omega_n^2}, \quad C = 2\left(1 - \frac{1}{\omega_n^2}\right)$$

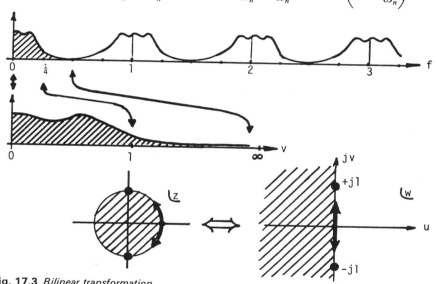

Fig. 17.3 *Bilinear transformation.*

17.4 Design of finite-impulse-response filters for general magnitude-versus-frequency specifications

The squared-magnitude of a non-recursive filter transfer function can be expressed in the form

$$\sum_{k=0}^{n} F_k \cos k\omega$$

for an nth order filter (which has $n + 1$ coefficients) This is because in polynomial $H(z) \cdot H(1/z)$ the coefficients of z^{+k} and z^{-k} are equal, and on the unit circle, $(z^{+k} + z^{-k})$ becomes $2\cos k\omega$.

The F_k coefficients are related to the filter coefficients h_k by

$$F_0 = \sum_{j=0}^{n} h_j^2 \qquad F_k = 2\sum_{j=0}^{n-k} h_j h_{j+k}, \qquad k = 1, n$$

where

$$H(z) = \sum_{k=0}^{n} h_k z^{-k}$$

Given a (periodic) magnitude vs. frequency specification, the F_k coefficients can thus be evaluated by Fourier methods; since the squared-magnitude is even and periodic, the cosine series is appropriate. Alternatively, the specification may be matched exactly at $n+1$ frequency points, giving a set of linear simultaneous equations from which the F_k coefficients are easily found. Non-linear optimization methods may be used to obtain an optimum (minimax) approximation.

The next step is less straightforward since although the F_k can be expressed explicitly in terms of the h_k, the converse is not true, and evaluating the latter from the former would involve the solution of a set of non-linear simultaneous equations. Instead $H(z) \cdot H(1/z)$ is obtained by using $2\cos k\omega = z^{+k} + z^{-k}$. The resulting polynomial has zeros in self-reciprocal pairs (e.g. if a is a zero so is $1/a$). After factorizing the polynomial, one zero can be selected (arbitrarily) from each pair, and these zeros used to form $H(z)$. It is usual to choose zeros within the unit circle, as this gives a minimum-delay transfer-function. The computational complexity lies in the factorization of $H(z) \cdot H(1/z)$.

If $H(z)$ is a self-reciprocal polynomial, the phase is exactly linear. (For such a polynomial, the coefficient array $h_0, h_1, \ldots h_n$ is a palindrome).

Thus,

$$H(z) = z^{-m} \sum_{k=-m}^{m} h_k z^{-k}, \text{ with } h_k = h_{-k}$$

for a $2m$th order filter, and

$$|H(j\omega)| = h_0 + 2\sum_{k=1}^{m} h_k \cos k\omega$$

since $|\exp(-jm\omega)| = 1$.

The magnitude of a linear-phase filter is thus identical in form to the magnitude-squared of a general non-recursive filter. The design procedure from a given magnitude versus frequency specification is therefore identical (i.e. Fourier approximation, solving linear equations and/or optimization) with the important difference that factorization is not required, since the filter coefficients are obtained directly.

Footnote The reciprocal of a polynomial, $P(x)$, is defined as $x^n P(1/x)$, where n is the polynomial degree.

17.5 Implementation

Digital implementation offers a direct trade-off between parallelism and pipe-lining to maximise speed, and multiplexing to minimise hardware.

The finite wordlength of the coefficients causes a discrepancy between the ideal and actual transfer function which is similar to the effects of component tolerances in analogue filters.

A fundamental difference in the digital case is that these effects are exactly predictable at the design stage and allowances can be made for them: there is no analogy with the random or ageing properties of analogue component tolerances.

The finite precision of arithmetic calculations generates a low-level noise signal (rounding-noise) which limits the dynamic range of a digital filter because of the need to maintain an adequate signal to rounding-noise ratio without arithmetic overflow. In practice, internal calculations generally need to be carried out to a precision of some 16 bits. Either greater precision or floating-point calculations would greatly ease difficulties which arise from the dynamic range limitation, but at present this usually implies an unacceptable cost or speed penalty.

17.5.1 *Implementation of recursive structures*
A software description of the second-order section of eqn. (17.1) is as follows:

```
loop
    TEMP := p * x1 + Q * x2 + K * u;
    u := TEMP + x1 + x1 + x2;
    x2 := x1;
    x1 := TEMP;
end loop;
```

The coefficients (P, Q. K) can be stored either in read only memory (ROM) or read write memory (RAM), while the state-variables must be in RAM.

To cascade m second-order sections, either the same routine (with different coefficients and state variables) may be run concurrently on m processors (with the output of one providing input to the next), or the routine may be run sequentially on a single processor (with the coefficients and state-variables of each section saved in RAM while the section is not being processed):

```
for i in 1 . . m loop
    load coefficients and state-variables for stage i;
    get input from stage (i − 1) output;
    process stage i, sending output to stage (i + 1);
    save state-variables for stage i;
end loop;
```

Note that the variable TEMP is local to each routine, and does not have to be saved.

150 Digital filters

Particularly if fixed-point arithmetic is used, the selection of section-ordering and scale factors are critically important aspects of obtaining the maximum dynamic range.

For a hardware implementation, it is possible to use a single fast multiplier, and arrange for coefficients and state-variables to be presented to it in the correct sequence by using a sufficiently complex controller. A compromise is a direct hardware implementation of the second-order section, using several separate multipliers, with multiplexing to achieve a cascade of second-order sections.

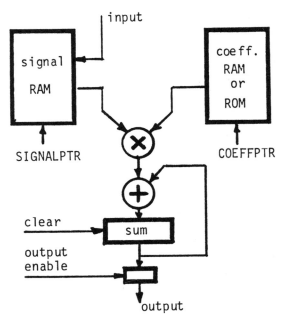

Fig. 17.4 *Multiplexing of a single multiplier-accumulator.*

17.5.2 *Implementation of non-recursive structures*

A direct digital implementation of the non-recursive filter-structure using separate multipliers and adders would be very costly. A single multiplier or multiplier-accumulator may be multiplexed as shown in Fig. 17.4. During each sampling-period coefficients and signal-samples are presented to the multiplier-accumulator in pairs, and after cycling through all pairs the output is enabled, a new input sample is stored in the signal RAM (overwriting the oldest sample), and the sum-register cleared. One of various alternative procedures is as follows:

loop
    sum := 0;
    index := filter-order + 1;

Digital filters   151

```
while index > 0 loop
   sum := sum + ⟨COEFFPTR⟩ * ⟨SIGNALPTR⟩
   increment SIGNALPTR modulo(filter-order + 1);
   increment COEFFPTR modulo(filter-order + 1)
   index := index - 1;
end loop;
output := sum;
⟨SIGNALPTR⟩ := input; -- overwrites oldest sample
increment SIGNALPTR modulo(filter-order + 1);
end loop;
```

Note that SIGNALPTR is incremented once more than COEFFPTR in each sampling period – this implements the 'shift' operation on all the signal samples without having to move their physical locations in memory. ⟨R⟩ denotes the location pointed to by R.

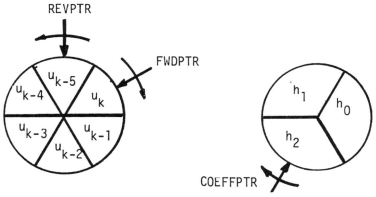

Fig. 17.5 *Address-pointers for 5th order linear-phase filter (shown in initialized positions)*

For this particular procedure, the sequencing of the inner (while) loop must multiply the oldest sample (μ_{k-n}) with h_n first, and the newest sample (μ_k) with h_0 last, and so COEFFPTR must be initialized to point to h_n.

The signal-samples of a linear phase filter may be paired to reduce the number of multiplications; thus, for a 5th order filter

$$y_k = h_2 * (u_k + u_{k-5}) + h_1 * (u_{k-1} + u_{k-4}) + h_0 * (u_{k-2} + u_{k-3})$$

The control-sequencing is more complex, and can be achieved by replacing the signal RAM address register (SIGNALPTR) by two pointers (FWDPTR and REVPTR) as illustrated in Fig. 17.5). A possible procedure for an odd order filter (e.g. for an even number of coefficients) is

```
FWDPTR := (REVPTR + 1) modulo(filter-order + 1);
```

152 Digital filters

```
loop
    MARK := REVPTR; -- save contents of REVPTR
    sum := 0;
    no-of-coefficients := (filter-order + 1)/2;
    index := no-of-coefficients
    while index > 0 loop
        sum := sum + ⟨COEFFPTR⟩ * {⟨FWDPTR⟩ + ⟨REVPTR⟩};
        increment COEFFPTR modulo(no-of-coefficients);
        increment FWDPTR modulo(filter-order + 1);
        decrement REVPTR modulo(filter-order + 1);
        index := index - 1;
    end loop;
    FWDPTR := MARK; -- reset to initial REVPTR
    ⟨FWDPTR⟩ := input; -- overwrites oldest sample
    REVPTR := (FWDPTR - 1) modulo(filter-order + 1);
end loop;
```

In all such schemes which involve multiplexing of a single multiplier, the execution of the outer loop must be completed within one sampling period if the filter is to operate in real time. A hardware or microprogrammed implementation allows various steps in the sequential descriptions of the procedures given above to be performed in parallel.

In addition to implementation entirely by software on a general-purpose digital computer, which places a severe limit on the maximum sampling-rate attainable, or a fully custom-designed chip, which is very expensive and lacks flexibility, the two principal approaches are to use either a microprogrammed bit-slice microprocessor or to directly implement a structure such as Fig. 17.4, using a fast bipolar or CMOS multiplier or multiplier-accumulator, together with a sequencer to generate the required timing and control signals. By using a writeable control store for the sequencer and RAM for the filter coefficients, fast but flexible programmable digital filters may be designed.

Single-chip 16-bit parallel multipliers and multiplier-accumulators are available with execution speeds of less than 100 ns, and single-chip floating-point processors are being announced which will offer comparable speed for floating-point multiplication or addition in a 32-bit (8 bit exponent, 24 bit mantissa) format. The availability of fast floating-point processor chips will substantially ease the implementation of digital filters, overcoming most of the problems of overflow, scaling, limited precision and dynamic range.

17.6 Conclusions

In a single chapter it is possible only to highlight a few aspects of digital filters. The topics included were selected in order to illustrate relationships between linear systems theory and digital filter structures, some of the frequency-domain design methods and the trade-offs available for implementation.

Chapter 18

Display types

Neil J. Bartlett
Plessey Displays Ltd., Station Road, Addlestone, Weybridge, Surrey, UK

18.1 Introduction

This chapter discusses those display types which are, or which potentially could be, used for graphics display applications within the fields of radar and sonar, and as text displays in these and other fields. In addition to the pre-eminent c.r.t., various flat panel technologies will be discussed – flat panel displays having a particular attraction in confined spaces and for mounting in desk positions.

18.1 *Common attributes*

Certain attributes are common to a number of display types, and to avoid repetition are discussed below.

Few display types retain their image permanently but require to be refreshed, this must normally be achieved at a minimum of 50–60 Hz if flicker is to be avoided. By the very nature of the refresh mechanism any point will only be energised for a small proportion of the time, the resultant duty cycle is significant since the smaller it is the greater the rate at which energy must be transferred in the refreshing process to maintain a given light output.

The essence of a matrix display is a set of row conductors and an orthogonal set of column conductors, with a display medium which is responsive to the sum of the drive applied to row and column, but not to either alone. By this means each pixel (as the points located at the intersections are usually called) may be addressed. Such matrix displays have a graphics as well as text capability. Most flat-panel display technologies have a duty cycle limit of around 1/300 due to the low energy nature of their drives, and employ line at a time refreshing to maximise the achievable display size. As an alternative to the graphics panels described most flat panel types can also be implemented in a dedicated text-only format in which conductors are not 'wasted' in the gaps between characters and lines. A minimum of 7 × 5 pixels are required to display English-type characters.

154 Display types

18.2 Cathode ray tube (c.r.t)

The c.r.t. comprises an evacuated funnel with a phosphor-coated screen at its wide end and the source of a focussed beam of electrons (gun) at the other. The beam may be deflected by a coil (yoke) at the neck, and modulated by a voltage applied to a grid in the gun.

The electrons are accelerated by an e.h.t. voltage typically 15 kV to 25 kV. The resultant high energy enables the c.r.t. to provide adequate brightness despite the very low duty cycle which is between 5×10^{-6} and 2.5×10^{-7}. The c.r.t. is used to display data in one of two ways, cursive or raster.

18.2.1 Cursive c.r.t. displays

Cursive is also known as caligraphic and this is very appropriate for the electron beam is steered across the screen in the manner of a pen or brush over paper (Fig. 18.1). The resultant continuous lines and positioning freedom differentiates the cursive c.r.t. from the other types of display to be discussed.

Fig. 18.1 *Cursive writing*

If the c.r.t. beam is made to track across the screen in synchronism with the radar or sonar receiver output signals, then the displayed image will yield the environment as seen by the sensor. In practice a modern display system will buffer the incoming signals in a store and play then to the c.r.t. in bursts. This retiming is not to be confused with scan conversion and creates 'spare time' during which the display can paint related computer-generated data.

Sensor data is essentially non repetitive and so to view and analyse it a long-persistence phosphor is employed. These phosphors are not bright and this constrains such displays to be used in dim lighting conditions, however refresh rates for computer-generated data as low as 25 or even 16 Hz may be used without flicker being seen.

Most cursive displays are monochrome, but colour can be provided. The normal colour system is beam penetration (penetron) and uses a special multi-layer phosphor whose colour is a function of the energy of the electrons and hence depends on the e.h.t. Penetron displays have the benefit of smooth continuous lines without any colour fringing, but the drawback of increased circuit complexity due both to the need to change the e.h.t. and to correct for consequent side-effects. Penetron phosphors also exist which change persistence instead of colour.

18.2.2 Raster c.r.t. displays

In a raster system the electron beam scans the screen in a regular pattern (Fig. 18.2). The beam is modulated by playing out the pixel-refresh store in synchronism. To achieve flicker-free presentation of the 1000 to 2000 lines needed for a high-resolution display requires a line-scanning frequency up to 120 kHz (seven times that of domestic TV), a video bandwidth up to 200 MHz and a refresh memory of up to 4 million pixels. Such a display represents the limits of current technology.

Fig. 18.2 *Raster-scan writing*

Sensor data requires to be processed in a scan converter (see Ch. 19), whence it will be refreshed on the display permitting viewing in bright conditions.

Colour can be provided by the shadow-mask technique though this requires that beam alignment is critical if colour fringing is to be minimised. Colour is not available in such high resolution as monochrome. Unlike penetron, shadow-mask gives a wide range of colours.

The c.r.t. is a large device and this can limit the freedom of the equipment or console designer, and in the cursive version the power consumption of the drive circuits is high. However with a grey-scale capability and high resolution, with a spot size typically one thousandth of the screen diameter, the c.r.t. is a versatile display device and the cursively written c.r.t. is especially convenient for the presentation of raw sensor data.

18.3 Vacuum fluorescent (v.f.d.) flat-panel displays

Like the c.r.t., the v.f.d. operates by bombarding a phosphor-coated anode with electrons, but the latter is constructed as a matrix, with multiple anodes forming one axis and grids the other. Cathode wires release an electron cloud and the whole is viewed by looking through the cathode and grid structure (Fig. 18.3).

The accelerating voltage for the electron is only 50–100 V which limits the duty cycle and hence matrix size. Interaction between the controlling elements

156 Display types

of adjacent rows/columns limits the pixel pitch, 0.8 mm being typical, though by increasing the complexity of the structure and its driving a pitch of 0.3 mm can be achieved.

V.f.d's up to about 320 × 240 pixels permitting the display of 1500 characters are available. v.f.d's are normally single colour, though colour can be provided by screening different phosphors to adjacent dots.

Character-only v.f.d's typically use a more complex internal structure that has dual grids, the benefit of which being the simplification of drive electronics.

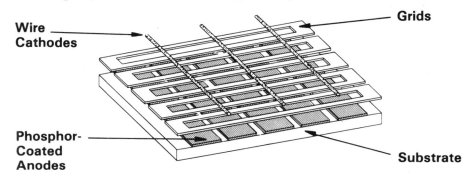

Fig. 18.3 *Vacuum fluorescent display*

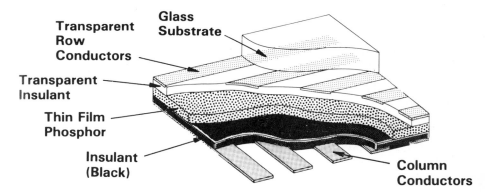

Fig. 18.4 *Sectional view of thin film a.c. e.l. panel*

18.4 Electroluminescent (e.l.) flat-panel display

Generating a high voltage stress by passing a current through a phosphor will also make it glow. This is the basis of e.l. and is typically implemented by providing about 100 V across some $2\,\mu m$ of phosphor. Two realisations are commercially available, a.c. thin film and d.c. powder.

The a.c. approach (Fig. 18.4) builds up on a glass substrate (through which

Display types 157

the display will be viewed) transparent row electrodes, an insulating layer, the phosphor layer, another insulating layer and finally the column electrodes. Current is capacitively coupled into the phosphor.

The d.c.-powder approach (Fig. 18.5) similarly uses a glass substrate and row and column electrodes, but has no insulating layers, and has a 30 µm thick layer of phosphor. This structure is electrically 'formed' to generate an active phosphor layer adjacent to the row conductors.

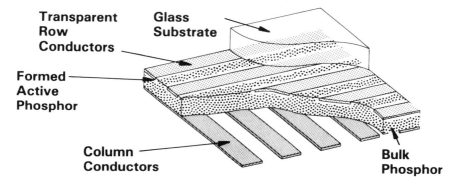

Fig. 18.5 *Sectional view of powder d.c. e.l. panel*

In either form, displays up to 2000 characters on matrixes of 340 × 250 pixels are available.

The power form has a white background and so incurs the same contrast ratio problem as the c.r.t. in that the ambient light illuminates the background. The thin film form has the potential to provide a black layer behind the active layer so improving contrast.

Potential exists to create multi-colour displays by stacking thin film structures each with a different colour phosphor, or by using a mixed phosphor whose constituent parts have differing intensities depending upon energisation period.

18.5 Gas discharge (plasma) flat-panel display

Plasma displays use the light emitted by a neon gas discharge, and in some versions use the difference between striking and sustaining voltage to provide memory. Both a.c. and d.c. implementations exist.

18.5.1 *The a.c. plasma panel*

The a.c. panel (Fig. 18.6) comprises a sandwich of neon gas between glass plates which have row and column conductors coated with an insulating layer. The application of voltage across all the conductors generates a field across the neon gas. Normally this is insufficient to strike a discharge. If at one pixel the voltages

158 Display types

are raised then a discharge will strike and current flow. This current builds up a charge on the cell walls. On the next half cycle, this stored charge is additive to the applied field and again causes a discharge to strike. Thus the display has the feature of inherent memory and does not require to be refreshed externally. The a.c. plasma display has thus broken clear of the duty-cycle and refresh-rate problem. Two million pixels are available in panels of 1 m size. The penalty of inherent memory is loss of a grey-scale capability.

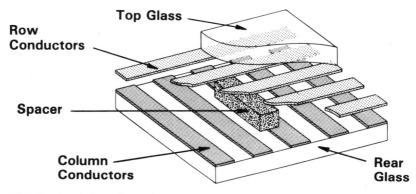

Fig. 18.6 *Sectional view of a.c. plasma panel*

Typical pixel pitch is 0.4 mm though 0.25 mm is possible. The structure of the panel means it is transparent, thus plasma panels can be overlayed on a map, or have data rear projected. The colour of the display is the characteristic neon orange. Alternatively it is possible to introduce phosphors which will glow with other colours, multi-colour displays can be made by placing different colour phosphors at adjacent pixel positions.

A.C. plasma is one of the more expensive display technologies, which tends to limit its application to where its unique advantages are necessary. For a long while a.c. plasma was the only flat-panel display with a capacity of 2000 characters.

18.5.2 The d.c. gas discharge panel
D.c. panels (Fig. 18.7) similarly have a glass neon sandwich with row and column conductors, but no insulating layer. d.c. panels additionally have a perforated spacer in the neon region so as to contain each pixel discharge. Matrix addressing occurs in the usual way and a direct current flows. Such a d.c. panel has no memory and so requires refreshing. Again colour can be provided by the use of phosphors. Due to the structure of the panel, dot pitch is quite large (0.8 mm) and multi-colour displays displays are not sensible.

Displays of up to about 500 characters are available.

An inherent memory version of the d.c. panel can be made by incorporating current control resistors at every pixel and providing a continuous maintenance voltage. Such displays are aimed at applications requiring very high brightnesses.

The consistency and rapidity of striking a neon discharge in isolation is inadequate for satisfactory panel operation and for this reason sustainer (or keep-alive) discharges are maintained at points which will not be seen by the viewer.

A number of derivatives of the basic gas discharge panel are possible, these include the use of rear-mounted keep-alive electrodes to scan the display to simplify it's addressing, drawing electrons off the discharge and accelerating them towards a phosphor-coated screen, and the combination of a.c. and d.c. techniques.

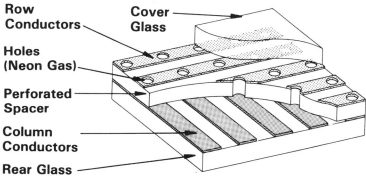

Fig. 18.7 *Sectional view of d.c. gas discharge panel*

18.6 Liquid crystal display (l.c.d.)

Unlike the previous types, the l.c.d. does not emit light but is reflective (or transmissive) and so can be used in very high ambient light levels. The construction of the l.c.d. is again a glass and conductor sandwich, this time filled with a 10 μm layer of liquid. The term liquid crystal is very appropriate for in its normal state the molecules of this liquid will align in a regular manner (Fig. 18.8a), but under the influence of a voltage field will realign along the voltage gradient (Fig. 18.8b). This molecular realignment is the basis of the light valve action of the l.c.d., different types of l.c.d. utilising the effect in different ways. The two most common types are t.n. (twisted nematic) in which a 90° rotation of polarised light (as illustrated in Fig. 18.8) is made visible with crossed polarisers, and g.h. (guest host) in which the molecules are dyed and either parallel with the direction of view (transparent) or perpendicular to it (opaque).

The multiplex ratio is severely limited since l.c.d's are not dependent upon the instantaneous, but upon the r.m.s., voltage. Larger panels can be made by using a complex topology which, for example, wraps round a 64 × 960 pixel display to look like a 128 × 480.

160   Display types

L.c.d's have a limited temperature range, but have the benefit of very low power consumption. There are few limits on pixel resolution and displays with a pitch of 0.2 mm are available. Multi-colour displays can be made by aligning coloured filters with the pixels.

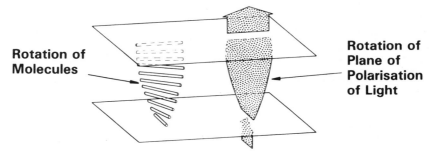

Fig. 18.8a *unenergised l.c.d.*

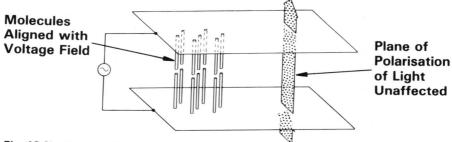

Fig. 18.8b *Energised l.c.d.*

18.7 The future

Especially in the realm of flat-panel displays, technology is advancing rapidly and it is virtually inevitable that some of the statements of capability made in this chapter will have been made obsolete even while the book is being printed.

The commercial targets of good-quality flat colour TV screens, and A4-size high-definition text page displays will be met in one technology or another. Progress will be made in eliminating refresh and multiplex problems by providing local storage at each pixel of the display. It is this storage that allows the a.c. plasma panel to be made so large, there is prospect of employing a similar memory property in a.c. e.l. Comprehensive facilities will be provided by building the display direct onto a large integrated circuit. l.c.d. and v.f.d. implemented on silicon slices have been made, but arguably the most exciting research is that directed to being able to manufacture integrated circuits deposited onto very large area substrates.

In all this, though, the c.r.t. must not be forgotten. Its demise has been predicted for years but c.r.t. system developments continue to occur at no less rate than in newer technologies.

Flat-panel developments can be expected to spin off into the fields covered by this book, initially as supportive displays and then increasingly as the sensor display though it will be a long while, if ever, that the c.r.t. is displaced as the prime high resolution display for sensor data.

18.8 Acknowledgements

The author gratefully acknowledges the cooperation of the Plessey Co. plc in the preparation of this chapter and for permission to reproduce drawings.

Chapter 19
Scan converters in sonar
J. W. R. Griffiths
University of Technology, Loughborough, UK

Three main types of system used in sonar to obtain a PPI or Sector type of 'acoustic image' underwater are:

(i) within-pulse scanning
(ii) mechanical scanning
(iii) sidescan sonar

In the first the whole sector or region is insonified by each pulse transmitted and the received beam is scanned very rapidly electronically across the sector at a rate which allows the receiver to 'look' in each direction at least once during a time equal to the transmitted pulse duration. A typical practical system (Ref. 1) operates at a frequency of 300 kHz, has a sector angle of 30 degrees, a receiver beamwidth of 1/3 degrees, a minimum pulse duration of 100 microsecs. and a scanning rate of 10 kHz.

A mechanically scanned system uses a narrow beam for both transmission and reception and the beam is stepped across the sector at a maximum rate of one beamwidth per transmitted pulse i.e., rather slowly unless the range is very short. A high resolution commercial system (Ref. 2) operates at a frequency of 500 kHz, has an angular resolution of 1.4 degrees, a range resolution of 80 mm and range scales up to 100 metres.

Finally the sidescan sonar. This is somewhat similar to the mechanically scanned system except that scanning is achieved by moving the sonar bodily through the water, usually by towing the system behind the ship, while the beam points broadside to the direction of motion. In this way a swarf is swept out on one side of the ship. By using what is in effect two systems it is possible to view on both sides of the ship at the same time. An example of a very large sidescan sonar developed by the Institute of Oceanographic Science is called Project Gloria (Ref. 3). This operates at a frequency of about 6.5 kHz, has a beamwidth of 2 degrees, a maximum range of about 20 km and a range resolution of 10 metres.

Typical outputs of these three systems are shown in Figs. 19.1–19.3.

Scan converters in sonar 163

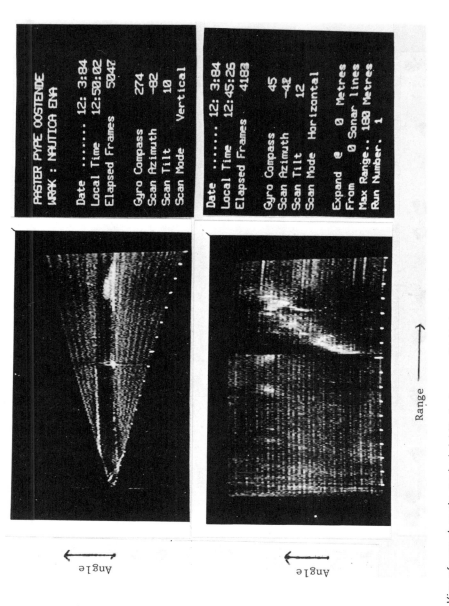

Fig. 19.1 *View of a wreck on the sea bed. (a) Vertical sweep. (b) Horizontal sweep*

164 Scan converters in sonar

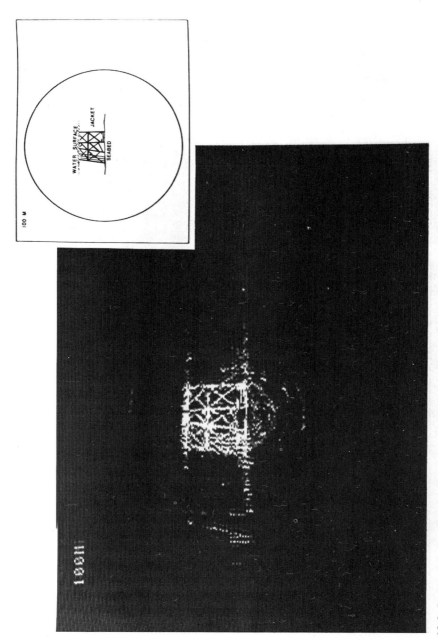

Fig. 19.2 *Profile of N. Sea Jacket taken with scanning head in horizontal position and close to jacket (after Low)*

Scan converters in sonar 165

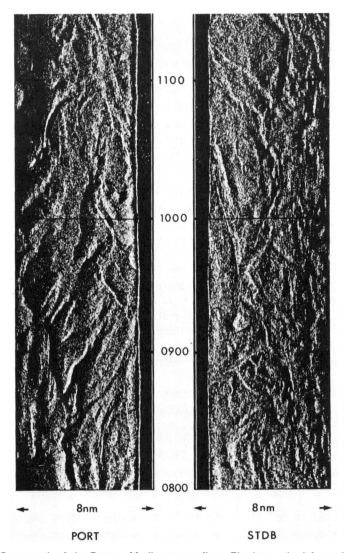

Fig. 19.3 *Sonograph of the Eastern Mediterranean floor. The intensely deformed soft sediments are shown well despite their relatively low relief. In these as in all sonographs the best visual effect is obtained by looking at them obliquely from maximum range* (After Somers)

The relative advantages and disadvantages of these systems have been discussed at length on many occasions; the discussion will not be repeated here since it is not the purpose of this paper, but a common problem, particularly of the first two types of systems lay, until recently, in the display of the data. For instance in the first system described above it was necessary to provide a raster type display with a line frequency of 10 kHz but with a frame frequency of only a few Hz, depending on the maximum range required. This necessitated a display with a very long persistence which normally had to be viewed in reduced lighting. The recording of the images was difficult and photography, with all its inherent problems, was nevertheless the main method used.

However, over the past decade or so both the price and the physical size of semiconductor memories have reduced enormously and it has now become possible to store a complete frame of a sonar image at an economical price. This stored data can be scanned simultaneously at the normal television rates so enabling the information to be displayed on standard TV monitors. Such a system brings many advantages.

(i) Brighter flicker-free images are obtained which can be viewed under normal lighting conditions. Additional, inexpensive displays can easily be provided at other locations, e.g., the bridge, which can help enormously in manoeuvring the vessel to optimise the sonar picture of an underwater structure.

(ii) Colour can be introduced on the display to represent amplitude, or for highlighting some particular feature or even just to make the display more pleasing to the eye. Since there are a very limited number of Just Noticeable Differences on a monochromatic display it would seem fairly obvious that the use of colour should increase the information transfer from machine to man. However, the use of colour as an information code has been the subject of extensive research and it is significant that opinion is still divided as to how effective it is in assisting the operator to process information. It seems certain that the benefits of colour are application specific and a decision as to whether or not to employ colour can only be made on the basis of practical experience. Suffice it to say that so far operators have expressed preference for the colour display for this application.

(iii) Images may be recorded using a standard TV video cassette recorder which, with the advent of the mass market for these recorders, have become very reasonable in cost.

(iv) The ability to put other information on the display besides the sonar images is very useful, particularly for alphanumeric data. This information is also recorded along with the sonar image.

(v) The display can also be synchronised to a computer display, such as that of a small microcomputer and the two displays can then be overlaid. Such an arrangement makes it very easy to add computer generated information to the display. Again the use of colour is an advantage. Diagrams can be added to facilitate the operator's rapid assimilation of the present state of some of the parameters of the sonar system, e.g. the position of the pan and tilt mechanism.

(vi) The flexibility of the display of the sonar image itself is increased. For instance it is relatively simple to provide an expanded part of the image as well as the full display and to offer alternate geometries of the display, e.g. rectangular or true sector display.

(vii) By mixing part of the old frame with the new frame as it is being stored it is possible to introduce controlled and variable persistence to the image. The equivalent circuit of such a system together with its effective impulse response is shown in Fig. 19.4. This process reduces the background noise on the display significantly at the expense of some slight smearing of the fast moving targets. By making some simplifying assumptions and assuming a feedback factor of β it can be shown (Ref. 4) that the gain in video signal/noise ratio is $10 \log ((1 + \beta)/(1 - \beta))$.

In practice the smearing does not appear to cause any problem providing the persistence is not too long. The fact that the amplitude is represented digitally allows the introduction of a feedback factor of 1/2 simply by dropping a digit of the old data before adding it to the new data. This gives a S/N improvement of about 5 dB.

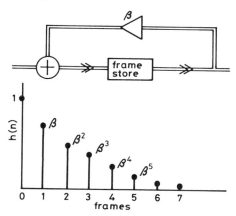

Fig. 19.4 *Persistence control*

(viii) The sonar image can be frozen on the screen if required in order to allow close examination or measurement of structure, e.g. wrecks, using overlaid computer generated cursors. If required the digitised image can be transferred to a computer for further processing.

(ix) In a sector scanning system, by using a frame store that is much larger than that required to store one frame of the sonar scan, it is possible to continue to store 'old' information which is not being currently updated and the system can be geographically stabilised. Thus a much wider sector, even the full 360 degrees, could be stored on the display. This is somewhat similar to the mechanically scanned system only the scanning is under the operator's control and he simply points the system in the direction of current interest. Furthermore it is possible to use electronic stabilisation of the display to allow for the movement of the ship although the determination of the appropriate coordinates is quite complicated (Ref. 5).

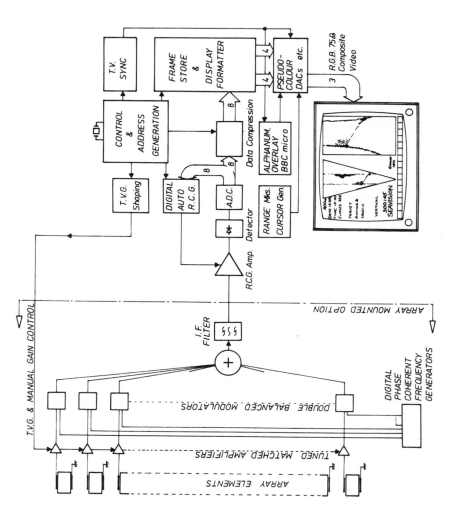

Fig. 19.5 Block diagram of Seavision

Scan converters in sonar 169

A block diagram of a small short range sonar using electronic scanning and scan converter display is shown in Fig. 19.5. The whole unit, including a small transmitter, fits into a 12U 19 inch rack module. A photograph of the display illustrating some of the features outlined above is shown in Fig. 19.6.

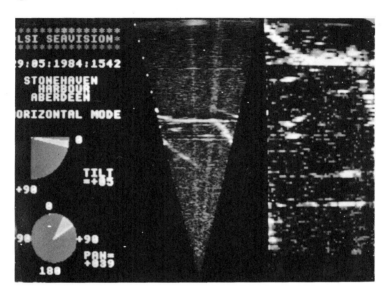

Fig. 19.6 View of 'Seavision' display

Acknowledgement

The author has pleasure in acknowledging the work of many of his colleagues which has been used in the preparation of this paper.

References

1. GRIFFITHS, J. W R., 1978, 'High Resolution Sonar System', Proc. Oceanology International '78, Brighton.
2. LOW, J. M., 1982, 'High Resolution Portable Scanning Sonar with Integral Digital Display', J.A.S.A. Suppl. 1, Vol. 72.
3. SOMERS, M. L., et al., 1978, 'Gloria II-An Improved Long Range Side Scan Sonar', Proc. Oceanology International '78, Brighton.
4. COOPER, D. C. and GRIFFITHS, J. W. R., 1961, 'Video Integration in Radar and Sonar Systems', J. Brit. I.R.E., Vol. 21, No. 5, May.
5. PHILLIPS, B., 1979, 'The Feasibility of Electronic Stabilising High Resolution Scanning Sonar Systems' I.o.A. Conference on "Progress in Sector Scanning Sonar" Lowestoft, Dec.

Chapter 20
Display ergonomics

V. David Hopkin

RAF Institute of Aviation Medicine, Farnborough, Hampshire, UK

20.1 Introduction

Ergonomics concerns the design of equipment for human use at work. With simple equipment it may be feasible to correct design deficiencies on the basis of operational experience, but in complex systems this approach is too costly, cumbersome, time-consuming, disruptive and ineffectual and it becomes essential to apply ergonomic principles during the specification and design of the system rather than to attempt to remedy ergonomic deficiencies after the system is in being. Modern systems with extensive software may seem to encourage once again system modification in response to operational experience, especially if the facility to modify in this way is claimed beguilingly to permit flexibility or adaptability, but such changes may lead to unwelcome human error unless the associated skills and learning can be transferred intact.

Ergonomics seeks an optimum match between many factors: operational requirements, equipment and facilities, human capabilities and limitations, tasks, physical environments and workspaces, and conditions of employment. Most ergonomic factors interact with each other. It is vital that no important ergonomic factor is totally ignored. Ergonomics is often applied in interdisciplinary contexts where numerous factions seek practical compromises. The practising ergonomist needs to know not only the facts and recommendations of his own discipline but the strength of the evidence for them, so that he can judge how far his recommendations may be compromised in the interests of interdisciplinary harmony and when they must not be.

Ergonomic recommendations are normally couched in general terms. Underlying presumptions may be that there is one best way to perform each function and that an optimum value or range for each dimension exists. Workspaces, and the displays within them, are designed to suit most prospective users rather than to be ideal for each individual. Only a few features, such as seat height or display brightness, are customarily adjustable to meet individual needs or wishes, and practical training or guidance is seldom given to allow each individual to make optimum adjustments. Optional calldown facilities, or interactive dialogues

Display ergonomics 171

where the machine adapts its level to that of the operator's own responses, are also designed for general use, and take account of individual differences as a means to that end.

20.2 Existing evidence

A considerable body of evidence on the simpler aspects of display ergonomics can be applied directly in most radar and sonar environments. Listed below are some factors on which existing ergonomic knowledge is normally adequate:

20.2.1 *The physical environment of the workspace*
This includes temperature (21°C or thereabouts); humidity (about 50%); air flow rate (0.2 metres/second); the colours (pastel), surface textures (matt), and sound deadening properties (acoustic tiles, carpeting) of surfaces; and ambient noise levels (60 dB maximum above 20 μPa).

20.2.2 *Workspace dimensions*
This covers room size; location of work positions within the room; accessibility; provision for amalgamation or sharing of work positions, for supervision and assistance, for on the job training, and for ancillary functions such as maintenance and cleaning; and anthropometric data about the body dimensions of the user population (NASA, 1).

20.2.3 *Seating*
This includes seat height and its adjustment (400 to 500 mm); seat depth (380 mm); back support; location of seat in relation to console, especially clearance between the seat and the underside of the shelf (200 mm); space under the console for legs (750 mm at floor level behind the vertical from the shelf front); and separation between adjacent seats (650 mm minimum seat centre to seat centre).

20.2.4 *Console profile*
This covers shelf height (700 mm); shelf depth (300 mm); shelf angle (between 0 to 10° from the horizontal); and numerous task dependent factors such as provision for controls and writing surfaces, the angle, size and relative positioning of main displays, and locations for controls and displays not in common use; console height; and the relative locations of console suites within the room (Figs. 20.1 and 2).

20.2.5 *Relations between displays and controls*
This includes viewing distance (normally 450 mm or thereabouts); reach distances (dependent on task, anthropometry and control characteristics); and recommended dimensions of characters and symbols on displays (dependent on

172   Display ergonomics

method of generation – for very clear characters about 3 mm character height for 450 mm viewing distance for tabular data).

20.2.6. Legal and medical criteria

These include radiation levels; minimum eyesight standards and acceptable forms of visual correction; renewal rates and consequent flicker (minimum 50 Hz to avoid flicker on most display types); and desirable postures, as determined by dimensions, seating, profiles, reach and viewing distances, and task characteristics.

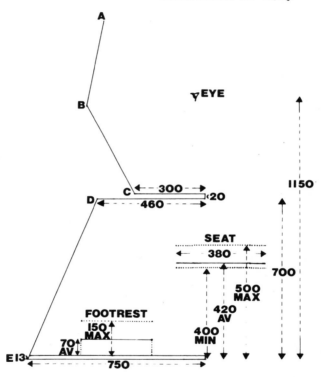

Fig. 20.1 *Example of console profile with large display.*

20.3 Ambient lighting

A noticeable omission from the above list is ambient lighting. In radar and sonar contexts there may not always be sufficient existing ergonomic evidence to specify the optimum ambient lighting from first principles. The problem is

Display ergonomics   173

interactions. Factors that should influence the spectrum and intensity of ambient lighting include the following:
The dimensions, layout, decor, colours and textures of the surfaces within the workspace;
The technical specification of available lighting;
The prevention of glares and reflections;
Display coatings and filters;
Display phosphors and renewal rates;
The resultant visual contrasts and brightnesses of the displayed information;
The tasks and the eye-movement patterns they impose;
The visual standards of the operators;
Work-rest cycles, especially the maximum duration of continuous periods of work (Colquhoun and Rutenfranz, 2).

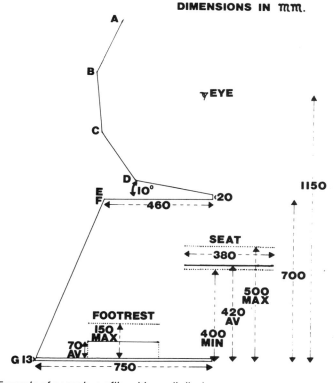

Fig. 20.2 *Example of console profile with small displays.*

Positioning the fitments can be critical, for example to bring console shelves near to office lighting standards (300 lux) yet with less light directly on radar or sonar displays and with no pools of light or darkness within the workspace. One objective is to minimise the need for frequent visual adjustments during task performance. Thus displays which must often be viewed in succession by the same operator should have approximately equal visual distances to prevent constant re-focusing, and approximately equal light output to prevent frequent changes in pupil size. Such design flaws may not actually impair performance much but they do engender complaints arising from subjective impressions of fatigue and eyestrain, and even though, because the eye is robust, they may not actually introduce an occupational health hazard, they are nevertheless best avoided.

With advances in display technology it is becoming more possible to optimise displays and lighting for human use by raising lighting levels and using coatings and filters to minimise unwanted interactions between the ambient lighting and the luminous contents of the display. However, in some environments, the lighting still has to be chosen to optimise the poor visibility of the display contents. As a result a period of adaptation may be necessary before the observer's eyesight has adjusted sufficiently to the lower lighting levels to function efficiently. In extreme cases some of the displayed information may not be perceived at all and other items frequently misinterpreted. Possible reasons are that the signal-to-noise ratio may be insufficient, the signal too weak, the background noise too prominent, the information too transient, inconsistent or faint, or real and spurious signals too similar visually to be distinguished reliably. In such circumstances, insufficient information may be present for the task to be performed to the required operational standards. If so, it is pointless to persist in attempts to select or train operators until they reach those standards. Even when a signal can be detected reliably, operational requirements may not be met, if the information is only sufficient to tell that there is something present but the operational requirement is to allocate the signal to a particular category.

20.4 Perceptual structuring

In both radar and sonar displays, interpretation relies on the structuring of visual elements into patterns. Where the data are relatively unprocessed, the perception of elements, and hence of patterns, depends on the ambient lighting, on the surface characteristics of the display such as filters, coatings and reflectance, and on adequate contrast ratio, defined as

$$\frac{\text{foreground} - \text{background}}{\text{background}}.$$

The interpretation of elements will depend on their perceived structure and visual groupings, and on their absolute values — elements of equal brightness may be judged as the same, or perceived as parts of a single visual structure. With such data, electronic enhancement may not be possible, and the problems of detection, discrimination and identification of information near the visual threshold are encountered.

Display ergonomics 175

In more processed or digitised displays, if the signal can be discriminated electronically from its background, it can usually in principle be amplified and presented in near-optimum form for the task. However, spurious signals may be treated in the same way, and be indistinguishable from real ones. Much information about patterns and textures may remain unprocessed and not appear. Thus from the point of view of the operator, display deficiencies and sources of human error tend to vary with display processing.

20.5 Visual coding

Electronically generated codings of information are used to present digitised data, or may be superimposed on unprocessed data. The main codings include brightness (2 levels), contrast inversion (one inverted level only), and size (3 levels maximum). Flashing is suitable to draw attention to a specific item for a short time, (always used sparingly, preferably not more than 2 rates, equal on-off cycles within each rate), but to draw an outline round an item often is a better alternative for longer times. Other possible dimensions, such as orientation, rotation and feature combination, can introduce unexpected errors: they are best eschewed unless already familiar, and always require practical verification.

Two major codings are shape and colour. Shape may be pictorial (i.e. representational), geometric, or alphanumeric. Criteria guiding choice should include visual discriminability from the background and from all other shapes, and unambiguous verbal labelling if display contents have to be reported verbally. Shape information which has to be discriminated, as distinct from information which is part of the background or merely adds visual structuring, must still meet contrast ratio requirements (6:1 to 10:1 against background). So must colour-coded information: this means that saturated reds and blues are often best avoided, and that certain saturated colours such as yellow may be too bright. High saturation is not generally necessary for effective colour coding, and pastels may give more acceptable contrasts. The contrasts of different colours with the background should not be identical.

Guidelines for choosing discriminable sets of colours are now evolving (Laycock, 3). Redundancy of coding is a sound practical principle whenever it can be introduced without ambiguity: this means that a given distinction is made independently in more than one coding dimension, in both shape and colour, for example.

20.6 Two specific problems

Numerous specific problems arise. One concerns alphanumeric labels on radar displays. Sets of alphanumeric characters can generally be designed from existing guidelines, but on radar displays the tendency has been for the label to contain more information to minimise cross referencing to tabular displays or other forms of information. This results in large labels which increase clutter and label overlap. Attempts to reduce the size of the labels make the individual alphanumeric characters too near the visual viewing threshold. The minimum character height for labels may be about 2 mm but depends on the way in which they are generated, on the contrast, the ambient lighting, and particularly the operational eyesight standards for operators. The poorest and most fatigued operator under the most adverse lighting conditions with the display near the end of its useful life must still be able to read the labels clearly.

On sonar displays, the reliability of interpretation is often a function of integrating the data visually into lines. Skill may reside in choice of amplification and selection of display frequency bands. Auditory and visual patterns may need to be compared to reach an interpretation of their joint meaning. If a verbalised discriminatory report is required, the conveying of information is not only a function of the successful visual discrimination of a pattern but also perhaps the comparison of it with a mental or physical library of patterns and the selection of a suitable word to denote it. A means to validate responses is essential to prevent the development of an agreed body of spurious knowledge which can arise if everyone is trained in the same way and responses are never checked independently.

20.7 The provision and use of information

This is an important distinction (Hopkin, 4). A task analysis should establish in principle what information is needed to perform a task to a required standard, but it is not sufficient to ensure that the information necessary for the task is present on the displays: it must also be useable. It should be clearly depicted, intelligible, unambiguous, compatible with the instructions for its use, correctly matched to the operator's knowledge and skills, at an appropriate level of detail for the task and for the time available to perform it, and portrayed in a way that is meaningful to the user rather than demonstrating a technological advance or innovation which may convey no benefits for the user.

It is easy to forget that the typical radar or sonar display is meaningless to someone with no specialist knowledge of it. Its contents are not self-evidently meaningful, nor does the display itself convey its purposes or indicate what tasks could be done with it. All the information required for all the tasks must either be displayed or be known to the operator through training and experience. In the latter instance, the system relies on human knowledge, on human memory, and on the operator's ability to recall the correct information whenever it is needed. Reliance on human memory in this way is fallible: it should be reinforced by displays wherever possible.

In the future, radar and sonar displays may incorporate more aids to problem solving, decision making, predictions and similar higher mental functions. These are likely to be associated with dialogues between man and machine in which displays have to present menus, indicate options, and ensure that dialogues have clear origins, direction, and conclusions, so that the operator cannot get lost. In such dialogues, one of the most difficult emerging problems is how the operator can discover quickly that certain information, expected to be present, is not in fact there.

Future displays may also need to convey other kinds of information: for example, they may denote the origins of data, what sensors are being employed, what quality control is being exercised, how accurate the information is, and how reliable. A good display should give some indication of how far the information on it should be trusted, and how it could be verified.

20.8 Failures and errors

Displays may need to show indications of technical faults and failures that are pertinent to task performance. Human adaptability to technical faults and failures depends partly on the provision made for relevant evidence to enable the operator to diagnose them and allow for them. This latter point requires good display-control relationships so that it is clear not only what to do but how to do it. It is an aspect of a broader issue: how much technical knowledge about the system does the user need in order to use it properly?

In designing systems, and choosing display contents and their codings, it is important to realise that the human errors that can be made and the confusions that can arise are largely predetermined by decisions taken during the system design and display specification. Technical innovations may introduce new uses and engender new kinds of error. Displays are the means by which the operator can acquire skills, experience, greater knowledge and greater competence at his tasks. They can only fulfil this function if they provide all the necessary information that he does not already know, and also sufficient information about the context of the tasks and about the outcome of his task performance so that he can learn and profit from his experience.

20.9 Measurement

Measurement of human efficiency using displays can be conducted at several levels (Hopkin, 5). System measures, comparing inputs to the whole system with outputs from it, indicate how efficient the system as a whole is, and how busy it is. Assessments of human behaviour such as general activity, task sharing, usage of communication channels, usage of keyboards, eye movement recording, and display viewing patterns show how the operator divides his time. Performance assessments, related to particular tasks and functions, indicate what he is achieving in terms of efficiency, speed, omissions, errors and proficiency. Subjective assessments obtain evidence on attitudes, thought processes, and discrepancies between what is actually achieved and what the operator believes he is achieving. Physiological and biochemical indices indicate the effort required to achieve the measured performance, and possible effects on well-being. Social measures indicate the role of teamwork, of official or unofficial social functions, such as supervision, assistance and task sharing, and of other influences such as peer pressure or group conformity in judgement. Biographical data may indicate relevant factors such as training and experience. Mathematically based descriptions and models may be used to describe the functioning of the system, of the displays, and of the operators, to predict the implications of proposed changes.

20.10 Résumé

Ergonomics, applied to displays, seeks to match operational requirements and tasks, human capabilities and limitations, equipment and facilities, and workspace and physical environment. Its aims are to promote efficient and safe task performance and the well-being of those who perform them. Almost all the parameters with which the ergonomist deals interact with each other: it is therefore particularly important that he does not omit anything of major significance. Ergonomics is normally applied to interdisciplinary environments which must often reconcile many disparate and perhaps conflicting requirements.

In general, the application of display ergonomic recommendations to sonar or radar contexts is best done through a human factors specialist. Display checklists, human factors guidelines and display manuals provide a great deal of basic data (McCormick and Saunders, 6). A further source is data originally compiled for particular contexts such as air traffic control (Hopkin, 7) or newspaper offices (Cakir *et al.*, 8), which require informed professional interpretation. The specialist will also know various journals that can be consulted for up-to-date data, and be able to judge the applicability of their contents to each human factors problem on radar or sonar displays.

20.11 References

1. NASA, 1978, 'Anthropometric Source Book', National Aeronautics and Space Administration, Houston, Texas, Report No. NASA RP-1024.
2. COLQUHOUN, W. P. and RUTENFRANZ, J., 1980, 'Studies of Shiftwork', Taylor and Francis, London, England.
3. LAYCOCK, J., 1984, *Displays*, **5**, 1, 3–14.
4. HOPKIN, V. D., 1976, 'The Provision and Use of Information on Air Traffic Control Displays', AGARD Conference Proceedings No. 188, 'Plans and Developments for Air Traffic Systems', Neuilly-sur-Seine, France; also in *The Controller*, **15**, 2, 18–22, and 3, 6–14.
5. HOPKIN, V. D., 1980, *Human Factors*, **22**, 5, 547–560.

6. McCORMICK, E. J and SAUNDERS, M. S., 1982, 'Human Factors in Engineering and Design', McGraw-Hill, New York.
7. HOPKIN, V. D., 1982, 'Human Factors in Air Traffic Control', Neuilly-sur-Seine, France, AGARDograph No. 275.
8. CAKIR, A., HART, D. J. and STEWART, T. F. M., 1980, 'Visual Display Terminals', Wiley, Chichester, England.

Copyright © Controller HMSO, London 1985

Chapter 21

Simulators

R. J. Morrow
British Aerospace, Dynamics Group, Bristol Division, UK

21.1 Introduction

Simulation, in one form or another, provides the equipment developer with the means of selecting the best design approach. The recent advances in the field of signal processing with the much improved processing electronics, software languages and development tools offer the promise of radically different solutions to a wide range of civil and defence requirements. Typically, the complex electronics based control systems that are at the heart of many avionics, plant control and weapon systems rely on the successful implementation of the numerous real-time processing tasks. Further, there is a general desire on the part of the customer for such equipment to have minimal human involvement in the operation and maintenance of the equipment. It is therefore evident that software based, multi processing, electronic designs will increasingly take on the burden of this much sought after capability.

As more of what has been regarded as human decision making is implemented in the machine, so it will become increasingly important to ensure that a full system performance evaluation has taken place. Further, it is equally important to have a means of design support to allow for progressive equipment performance updates and extension through the mechanism of software change. It is against this perceived need that the subject of simulators is treated. These simulators designed as total system development and validation tools, are proposed as a necessary pre-requisite to establishing successful advanced signal processing designs.

21.2 Simulators: aids to advanced signal processing

21.2.1 *Development, validation and post delivery support*

To realise the major benefits of the silicon chip and advanced software technologies; that of system redesign through software update, a system-level simulator support strategy is worthy of consideration during the feasibility study stage of a high-cost project. This strategy should be based on the long-lead development of simulators that may be used in a co-ordinated manner during all phases of product development. In this manner, the cost of the provision of a simulator may be amortized across the three major phases; as a tool for design use during development, as system performance providing facilities during customer acceptance and finally as a reference facility to support post-delivery improvements. A further possibility is the use of the same simulator, or a derivative of the design, to support operator training.

It is an unfortunate fact that the positive economies offered by microprocessor electronics are in many cases off-set by the high cost of software development and validation. Further, the roles of hardware and software design in achieving the desired system performance are highly inter-related. As such this has to be taken into account when assessing the benefits of using a generic simulator facility. The ability to include all aspects of the system performance in a total simulator plus equipment simulation will be considered in greater depth with practical examples.

21.2.2 Simulation, stimulation and emulation

The term 'simulator' is used to cover a range of techniques some of which are more appropriately defined as follows. In general, for complex or high-value systems, the designer will generate system models from which detailed hardware design specifications are derived. These models do not require hardware or direct equivalents of the hardware but represent the main aspects of the system. These are taken as examples of simulation.

When hardware exists it may be subjected to practical testing with meaningful test signals. If the objective of such testing is to verify that the hardware meets the desired system performance, then the test signals have to be injected into the hardware at the operationally correct data rates and the working frequency band. This approach is defined as STIMULATION and the method of modelling and creating the operationally correct test stimulus may be through the use of a simulator.

In the case where part only of the system exists as hardware, say, the processors but not the sensor, it is possible to model the action of the missing element and to produce a replica of the signals it will generate. As this replica must operate at practical infomation data rates, with an accurate representation of the hardware process, this is defined as an EMULATION of the missing hardware element.

21.2.3 Generic simulator elements

In order to define typical approaches and simulator elements the generalised signal processing system depicted in Fig. 21.1 shall be used. As shown, the control processor has a real-time control strategy based on the use of sensed analogue electromagnetic information, which is used to alter the use of the sensor(s) and elements of the physical environment referenced to the sensor. The

182 Simulators

information fed to the central signal processing unit is in digital form and the operation of the processor follows the software algorithms built into the equipment. Certain major parameters are displayed and kept updated by the processors to allow a human operator to interact in real-time with the system. The optimum approach to system proving is to create a high fidelity representation of the real world conditions to be met by the system during its normal operation.

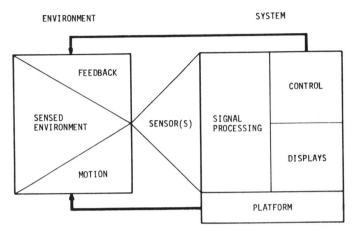

Fig. 21.1. *Generalised signal processing system*

21.2.3.1 *Operational environment*

Where developed hardware exists its performance may be evaluated through field trials. Unfortunately, increasingly this approach is not possible for reasons such as: safety limitations imposed by either man or machine, practical limits to realisable test environments, high cost of instrumented field trials, time scale, etc. An additional factor that has to be taken into account, when planning to use field trials for performance assessment, is its lateness in the equipment development cycle. Modifications, resulting from proving that full system performance has not been met, are very costly at this stage. Alternatively, field testing of the hardware at the prototype stage is not necessarily possible as it is not usually sufficiently robust or physically representative. Due to these reasons it is natural to propose simulation of the sensed environment to be the baseline simulator function.

21.2.3.2 *Motion*

If the sensor(s) are attached to a moving platform or the sensed environment has mobile elements these effects must be introduced as dynamic modification of the simulated environment. This may be achieved either by moving the equipment relative to the environment or by altering the simulated environment relative to the static sensor to provide simulated motion.

21.2.3.3 Feedback

A control system is depicted in Fig. 21.1 where the signal processors make continuous decisions to alter element in, or relative to, the sensed operational environment eg. modelling flight control or processing plant control.

21.2.3.4 Baseline elements

Together the three elements previously discussed may be combined to form a simulator that is able to stimulate the equipment via the practical sensor hardware. This may be developed such that it is available at the prototype stage and thus offer the developer performance test data at an early stage in the programme.

21.2.3.5 Sensor emulation

The above approach may be taken one step further by replacing the sensor which may have a longer development programme than that of the control electronics. Further, it is expected that in future advanced signal processing systems software development will be a major non-recurring cost element. For this reason, sensor emulations to allow early design evaluation of the processor control software can prove economic. The general availability of processors and standardisation of software languages supports this approach.

These generic simulator elements have been discussed with reference to a generalised problem as stated in figure 21.1. In amplification the following practical simulator examples have been included.

21.3 Dynamic electromagnetic environment simulators

21.3.1 Introduction

Two examples of comprehensive simulators designed to meet the approaches discussed in the previous section have been selected. The first demonstrates the extent to which a simulator can replace traditional field testing of a complex system; a missile, with a totally in-door trials facility. The second cites the use of the same signal processing technology that is currently being introduced in advanced control equipment, to provide a solution to a complex simulator design problem.

21.3.2 In door EM trials ranges for system performance evaluation

Missile guidance design has become too complex to be adequately developed and evaluated by reliance simply on analytical simulations and sparse flight test programmes. A full in-door electromagnetic trials range based on total environment and motion simulation with hardware in the loop provides an alternative approach. The general structure of the missile simulation facility is shown in Figure 21.2. The equipment comprises of 3-axis rate table, master hybrid

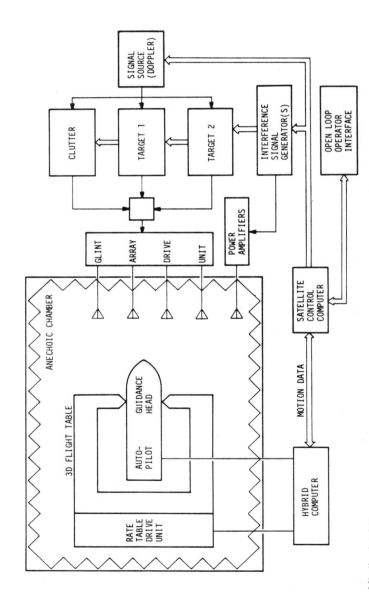

Fig. 21.2. *Missile simulation*

computer, target-generation satellite computer, RF target-signature generation equipment and antenna array with anechoic chamber.

In this configuration the missile is placed on the 3-axis table and its pointing direction with respect to the target antenna array continuously and automatically controlled from the master hybrid computer. In this manner the rate table provides dynamic rotational pitch, roll and yaw forces on the seeker. The guidance loop is then closed through the autopilot and control-systems kinematics model resident in the master hybrid computer.

Since the problem of analytically modelling the complex seeker head is overcome through the incorporation of the practical equipment, it becomes important to synthesise the RF environment as realistically as possible.

In the case of the missile sensor this envoronment will include a combination of wanted and unwanted signal returns produced from the target and the physical environment surrounding the target. The generation sub-system shown includes the satellite computer, electromagnetic signal generation equipment and antenna array. The input to the sub-system is the target and environment position change as digital descriptors. The output from the sub-system is the composite set of electromagnetic signals that are propagated via the antenna array to stimulate the missile seeker head. In stimulating the seeker head in this manner the control of the seeker received power level adds the fourth dimension of motion freedom: range to target.

This approach to missile environment simulation with hardware accurately stimulated at practical operating frequencies is equally applicable to both radio and electro-optical missile-seeker sensors. The principle equipment difference in the electro-optical case is the method of optically stimulating the sensor head. In practice this is implemented as an optical transmission path equivalent of the RF signal-generation equipment, RF antenna array and anechoic-chamber propagation paths. A more detailed treatment of the design of this simulator has previously been published, (Morrow 1).

21.3.3 *Multiprocessor based radar environment simulator*

Modern military operations on land, at sea, or in the air are characterised by their extensive reliance on electronic and radio equipment and naturally this usage has led to the development of a wide range of electronic reconnaissance aids. These reconnaissance aids are generally referred to as Electronic Support Measures, ESM and include those equipments that are used as monitors of the electromagnetic spectrum. Their purpose is to detect, locate and classify all hostile or potentially hostile platforms from prior knowledge of typical types of known hostile related emissions such as radars.

The dense multiple radar signal environments to be observed during combat are not available in peace time. For this reason only simulators can create representative electromagnetic spectrum usage. The objective of the emitter simulator is to recreate the original environment presented to the ESM sensor as, say, fitted to an aircraft or ship. The major simulator design aim is to

186 Simulators

reproduce all the characteristics of each of the transmitted signals in the environment.

The nature of an ESM radio receiving system is to observe dense emitter environments containing tens of hundreds of emitting sources. Typical interleaved pulse rates in excess of one million are common. The basis of modelling is by simulating each individual radar emission and then to time and frequency multiplex these many elemental emissions. A digital representation is created and used to control RF signal generators that directly interface to the equipment under evaluation, (Morrow 2). This is an identical approach to that previously discussed for missile simulation.

The design is approached from an advantageous proportioning of a real time digital drive to programmable RF signal generators. The necessary fidelity and flexibility for emitter representation, is obtained using a design based on the continuous solution of the real time processing task, using a special to type multiprocessor and high level software programming.

The extent of the real-time processing task may be grasped from the input and output data rates to and from the multiprocessor. The input to the multiprocessor are data updates defining the ESM platform's position and heading typically, 256-millisecond updates. The multiprocessor output has to define a digital descriptor for each interleaved pulse in the environment, a minimum of one million per second. Each emitter descriptor will include digital-word descriptions of the value of variables such as: time, carrier frequency, amplitude, pulse width, bearing, polarisation, intra-pulse modulation. To provide this type of flexibility upward of 64 bit fields are needed for each descriptor.

To meet this processing task the multiprocessor is constructed as a set of parallel processing data generation channels, as shown in Figure 21.3. Each processing channel is used to model one or more radar emitters. In practice this is realised as a modular hardware construction where the size of the processing unit may be built to match the desired size of the synthesized environment. Each processing channel is linked to a central control computer. This provides the interface for the real-time motion updates and allows each parallel processing channel to be loaded with software emitter model(s) which run as autonomous parallel processing tasks.

Software emitter models are table driven as equation solving is too slow. Each of the basic tables of emitter attributes is computed prior to a simulation. Off-line high-level-language-based (CORAL) routines are used to incorporate the flexibility and fidelity of mathematical modelling. The upper limit to fidelity is a function of the maximum processing speed of each elemental unit and the size of memory each may directly access. This memory may range from 16 K bytes to 64 K bytes. The 'on line' real time programs are written in high-level language (CORAL) for both the executive computer and cross compiled for loading to the parallel processing units. The emitter simulation program operates as a foreground/background process. The background processing relates to slowly varying parameters i.e. bearing, range and screening. The foreground processing is based on table sorting and emission description output.

Simulators 187

Having described the approach to real-time processing there remains a design problem of how to create a time ordered set of emission descriptions. The solution is to initialise all processing units from a master timing signal. From this event each processing unit computes emitter description time histories for selected time periods. These data are output to buffer stores; individual buffer stores are polled by the output interface. To each emission description a time stamp is attached. This defines time in 0.1 μsec steps for the whole of the selected time period. The emission descriptions are written to a sequence store memory using the time stamp to address the memory. The memory is sufficiently large to hold all the time period descriptor data.

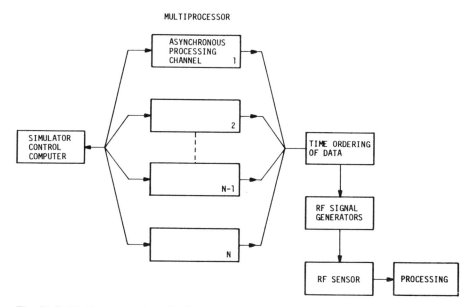

Fig. 21.3. *Multiprocessor based radar environment simulator*

The emission descriptions are passed via the interface to associated RF generation assemblies where individual parameter words are used to set associated functional items of RF processing hardware.

This type of flexible approach to the simulator design is only recently practical due to the rapid advances of microprocessor electronics and software language aids such as cross compilers.

21.4 Field portable trials simulators

21.4.1 *Introduction*

The previous section presented examples of simulators developed to meet a strategy of total simulator support including hardware totally integral with the

188 Simulators

simulator. A major disadvantage sometimes cited against this strategy is that aspects of the environment are not known in sufficient detail to ensure accurate system performance validation. In these cases it is often necessary to perform supplementary field trials as a validation of the total simulation. Here again the same simulator, or a derivative, may be used. An example of such a simulator for field trials with radar is given in the following section. The general approach is to include the general environment as a practical input but to have other system parameters as variables obtained through simulation.

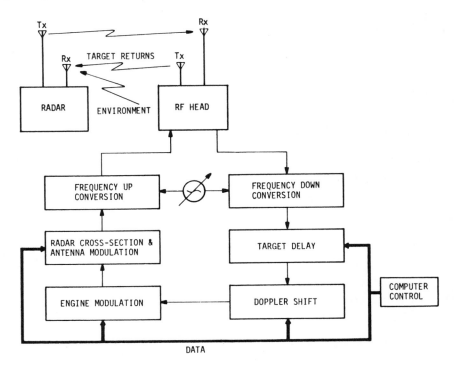

Fig. 21.4. *Transponder radar target return simulator*

21.4.2 Transponder radar target return simulator

The problem presented is that of evaluating radar designs. The radar's environment is determined through active electromagnetic interrogation. As such the radar is the RF source and the environment is the modulator of the signals back scattered to the radar's receiver. This is an important fact to be employed in the simulator design.

The major characteristics of radar electromagnetic environment synthesis is the treatment of the back scattering. Two classes of scattering are defined as, targets which are objects for detection, tracking or interrogation and clutter which encompasses any other scattered return. As such, clutter may be regarded as those signals that degrade perfect signal processing. Any object may be classified as either target or clutter depending on the function of the radar.

Radar resolution is the dominant factor that influences the approach to the simulator design. Objects that are physically small compared to the radar's resolution cell (i.e. that unambiguous time and space, field of view of the radar) can be treated as point scatterers. Objects that are large have to be treated as distributed or extended scatterers.

Following the same approach, as described in the previous sections, a simulator may be designed to produce a full representation of the environment, such that it may be used to directly stimulate the radar receiver via cable coupling behind the antenna. For point scatterers this is a practical approach but extended scatterers present a problem of scale if a deterministic testing approach is adopted. Thus in summary, extended scattering is computationally expensive to implement and an alternative approach is chosen, Wall (3).

The alternative approach is to use the fact that the environment is seen as a time delayed response to the radar's transmission, which if a simulator is designed as a transponder, it may be employed as a field instrument. In this manner the computationally expensive clutter is physically represented by the actual returns and the target is kept as a simulation. This allows target characteristic variation under scientific control with known conditions of clutter, as shown in Fig. 21.4.

21.5 Concluding remarks

The rapid advance of software based processing systems has encouraged the replacement of human control and decision making by machines. Where this involves high cost, human safety or use in extreme environments the system performance has to be rigorously established. In these cases simulators will be needed and their planned use to support multiple functions during the life of the equipment is sensible and economic.

In an equivalent manner the availablilty of advanced signal processing techniques will benefit the simulator designer. Future simulators will be increasingly capable of reproducing high-fidelity representations of practical conditions. These facilities will complement the rate of advance of the prime equipment.

21.6 Acknowledgements

The author wishes to acknowledge that the simulators described are products of British Aerospace, Dynamics Group, Bristol Division, and as such are the work of a number of the author's colleagues.

The author would also like to thank the Directors of British Aerospace, Dynamics Group, Bristol Division, for permission to publish this paper.

21.7 References

1. MORROW, R. J. 'Simulators for Electromagnetic Environment Synthesis' IEE International Conference on Simulators, Sep. 1983. Conference Publication No. 226. p. 19–24.
2. MORROW, R. J. 'Electromagnetic Environment Synthesis – A Tool for ESM System Development and Support'. IEE Proceedings Part F: Special Issue on ESM, 1985.
3. WALL, J. A. H. 'Target Simulator for Frequency Agile Radars'. Proceedings European Microwave Conference 1981.

Chapter 22
High throughput sonar processors

T. Curtis
Admiralty Research Establishment, Portland, Dorset, UK

22.1 Introduction

Current generation sonar acoustic signal processing systems operate at throughputs comparable to those required in fifth generation computer applications; for example, sonar systems are currently deployed with processing throughputs in excess of 100 million operations per second: these processing speeds are achieved using distributed networks of hardware based primitives or macros that can be programmed to execute the basic signal processing function. The networks are configured by simple control software to implement complete signal processing flow graphs that define the entire sonar processing function.

The performance of the next generation of operational sonars must be improved to counter the effects of target noise reduction and anechoic coatings. Improved performance dictates the use of large-area sensor arrays, with many more processing channels and "smarter" processing algorithms such as adaptive or high-resolution processing at the front end of the system and image processing methods at the display end. In addition, operational systems require fault tolerance, comprehensive online monitoring and maintenance facilities, reconfigurability and improved development aids. An increase in processing throughput of two or three orders of magnitude over current systems is necessary to provide these functions.

Advances in semiconductor and integrated-circuit technology promise significant improvements in device performance over the next decade, but it is unlikely that these alone will provide the necessary increase in systems performance. Comparable improvements are required in the areas of algorithm development and system architecture if the capabilities of the advanced technologies currently being developed in various national VHPIC and VHSIC programmes are to be fully exploited.

The following sections briefly review some aspects of recent developments in these fields and their impact on digital signal processing systems in future sonar applications.

22.2 Technology background

Digital signal processing methods have been used in sonar systems since the late 1950's (1), although these early processing schemes usually employed fairly crude signal-processing algorithms and were limited in performance by the technology available for their implementation (2). One of the major technology problems in early systems was the difficulty in storing the large amounts of digital data required for processing, particularly for example in conventional beamforming systems.

Many techniques were pressed into operation in an attempt to provide this storage function and the early history of the application of digital signal processing in sonar follows closely the development of computer mainframe storage technologies. Early systems used lumped L-C, wire, quartz and ZTC glass delay lines as well as magnetic core for data storage.

The first logic families (3) became available in the early 1960's and were widely applied in many digital signal processing applications. Even then, many processors used hard-clipped data to minimise hardware complexity. For example, digital multi-beam steering systems, e.g. DIMUS (4), clipped the hydrophone data and used clocked shift-registers to provide single bit, multi-tap delay lines. Using hard clipped data produced quantization and capture effects (5) which degraded system performance and multilevel systems to reduce these problems were developed in the mid/late 60's using early SSI MOST dynamic shift registers (6). Multilevel quantization systems provided improved performance but were constrained to use simple architectures by the sequential address accessing inherent in shift register based systems.

Random access memory (RAM) stores were developed in the late 1960's, when more complex processing architectures could be considered and compact processing systems developed. At that time, the devices available for arithmetic processing were of limited capability and only simple non-programmable systems could be considered, but during the following decade, high density RAM, read only memory (ROM), arithmetic/logic units (ALU's) and microprocessors became available. These allowed programmable systems to be developed for conventional sonar processing applications (7).

Currently a wide spectrum of LSI devices, for example high-speed arithmetic processors, single-chip micro-computers and high-density RAM, are available (8) and these provide sufficient processing power for complex conventional and small real-time adaptive systems to be built.

Over the past few years, various national and international technology programmes have been put into place (e.g. the US VHSIC, the UK VHPIC, the EEC ESPRIT programmes, etc) which aim to provide the enabling technologies and CAD design capability for the next generation of semiconductor devices. These programmes are now starting to provide the design tools and techniques that will enable ultra large scale integration (ULSI) on silicon (9). Techniques such as wafer-scale integration, coupled with advanced layer processing, multi-

High throughput sonar processors 193

level metallisation and sub-micron geometries will eventually allow complete systems to be implemented on silicon.

At the same time, new materials technologies are providing improved and novel semiconductor substrates, for example gallium arsenide, MBE superlattice structures, heterojunction devices, etc, with enhanced intrinsic semiconductor properties which will allow device parameters to be matched more closely to specific system requirements.

The emergence of these advanced technologies over the next decade and, perhaps more importantly, the widespread availability of the CAD design tools being developed to support them, will require a fundamental review of the techniques that have been employed in the past to develop signal-processing systems, if full advantage is to be made of technology improvements. In particular, existing processing algorithms must be "massaged" and new algorithms developed that match the emerging technologies and architectures developed to provide high-throughput processing systems for future generation applications.

22.3 System requirements

In a very general sense, the basic algorithms used in sonar systems are similar to those employed in radar or telecommunications and are well reported in the open literature (10, 11). Whilst the difference in the propagation physics of acoustic signals in the ocean and electromagnetic waves in space would at first sight indicate wide differences in the processing bandwidths needed for sonar and radar or communications systems, in practice this is not the case: the operational requirements to provide long-range detection and all-round surveillance cover in sonar processors requires systems with overall throughput requirements at least comparable to and often in excess of those used in other digital signal-processing applications. However, the detailed requirements for sonar digital signal processing are often very different from those developed for other applications. Simultaneous surveillance cover over a wide field, combined with good detection and localisation performance, dictates the use of large sensor arrays and multi-channel processing techniques and a number of different detection processes must be provided simultaneously to ensure a high probability of detecting a variety of targets with different characteristics over a range of operational conditions.

Typically, for a passive system, broadband correlation, energy detection, narrowband spectral analysis, vernier detection, DEMON analysis, transient analysis, etc, are all required simultaneously for target detection, classification and tracking. Even using relatively simple processing algorithms, the processor throughput required to implement these systems is daunting, typically in excess of 100 million operations per second (MOPs): Table 22.1 indicates the processing power required for the various operations in such systems.

In more realistic applications, with large area arrays and more sophisticated receive processing, these figures can be increased by several orders of magnitude. Consequently, although the per-channel bandwidth on sonar processing systems may be low, the large number of processing channels and the need to provide a range of different detection and localisation processes simultaneously requires systems with very high throughputs, typically in the range 10^3–10^4 MOPS. If these figures are compared with the throughput available from a number of commercial processors running signal-processing tasks, see Ref. 12, it can be seen that achieving this level of throughput is not trivial.

Table 22.1 *Typical small system throughputs*

ACTIVE SYSTEM
64 stave array, 0–10 kHz
64 beams
LPFM/CW transmission
Replica Correlation and
Doppler Processing
Broad band
BEAMFORMING – 48 MOP/sec
SIGNAL PROCESSING – 32 MOP/sec
PASSIVE SYSTEM
256 element array, 0–10 kHz
256 beams
Narrow band Surveillance and
Vernier Processing
Broad band, DEMON, etc.
BEAMFORMING – 40 MOP/sec
SIGNAL PROCESSING – 35 MOP/sec
1 MOP/sec = 10^6 operations/second
each operation of form-
$\{A * B + C\} \to C$
+ concurrent address generation, data accessing, etc.

In the past, high-throughput systems have been implemented using either dedicated custom hardware processors or programmable software-based multiple array processor systems. Both of these approaches have inherent advantages and disadvantages. Dedicated hardware processors achieve very high throughputs and wide bus transfer bandwidths, but lack flexibility and are difficult to modify or reconfigure to meet changing operational needs, or for other applications. Hardware systems often use many different card types and are costly to design and support, especially in military applications.

Software-based systems are programmable and hence more flexible and often require only a limited number of card types: however, the very fact that the system can be re-programmable can be used as an excuse to avoid making the necessarily difficult design decisions at the correct stage in system development. This has resulted in systems with excessive software complexity and more options than operators can usefully employ and represents misuse of system programmability. Some of the common consequences of this include under-utilised arithmetic units, "control bottlenecks", limited parallelism, poor performance prediction, high integration costs, monolithic system software and results in increased system costs.

In consequence, it has been difficult in the past to develop high throughput signal processors using either totally hardware- or totally software-based technologies. Both approaches have required long and costly development programmes and have often produced systems with limited flexibility and strength potential.

Processing schemes with architectures mid-way between the dedicated hardware and totally software-based systems are currently being developed to resolve these problems (13, 14). These new approaches aim to reduce system development timescales and costs by exploiting advanced technology to provide sufficient flexibility and processing power for next generation applications.

22.4 System architectures

For future generation systems, the single arithmetic unit processing architectures (e.g. vector processors) now used in many digital signal-processing applications run out of steam. Throughput can be increased to some extent using arrays of vector processors, but such an approach depends upon the development partitioned algorithms and smart schedulers to supervise algorithm scheduling and maintain concurrency. The higher the throughput required from a system, the more difficult the scheduling problem becomes. Processor scheduling is easier in applications where the required functions either have a high degree of granularity, when the problem can be decomposed into well defined individual task and implemented using highly parallel collections of autonomous MIMD (multiple instruction, multiple data) processors, or when the system function required is sufficiently well-ordered to allow it to be mapped onto a regular array of identical closely-coupled processors, as for example in systolic processing schemes (15) or SIMD (single instruction, multiple data) distributed processing systems (16).

Most sonar signal-processing problems are not well structured but are granular at the system level. They often require "long-range" interactions between various subsystems and consequently a flow graph network architecture (17), which supports these interactions, is more attractive than regularly structured architectures with only "short-range" interconnection paths. In many applications, the overall system function can be assembled from a collection or network of well defined sub-functions (or primitives); for example, many sonars can be built up of blocks performing, say, transform processing, matrix manipulation, filtering, beamforming, normalisation, display, etc. This allows a relatively straightforward mapping from a conventional system block schematic to a flow graph describing the overall sonar function in signal processing graph notation (SPGN) (18).

In the longer term, a direct mapping from system specification to system configuration may be possible, ie generalised functional decomposition into SPGN.

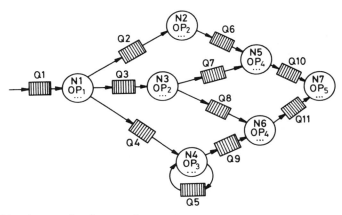

Fig. 22.1 *Signal processing flow graph*

To achieve high throughputs, SPGN systems are being developed which are data-flow driven (14): these systems allow control and scheduling functions to be distributed across the processor network and separate the high-bandwidth data paths in the processor from the slower control and monitoring highways. Figure 22.1 shows a typical signal processing flow graph: it consists of a collection or network of nodes which perform the basic signal-processing primitives, interconnected by queues. This flow graph describes the overall system signal-processing function when operations are assigned to the nodes. Hence the graph represents the function decomposed into its basic activities and their interconnection into a system. In this sense, the signal flow graph is analagous to an activity diagram in a MASCOT system and can be used in much the same way to define individual node processes or groups of node processes (sub-graphs or clusters) with well defined control parameters and input/output transfers via queues. The graph is overlaid with control and monitoring networks which allow the node primitive parameters and the data paths through the network to be modified.

High throughput sonar processors 197

The basic structure of the primitive used in one such signal processing flow graph system (19), shown schematically in Figure 22.2, contains four main highways: high-speed parallel input and output signal flow paths and separate control and monitoring highways. The primitive contains the local timing, microcode generation and control circuits, as well as the input queue FIFO store and the arithmetic unit which performs the actual macro calculations.

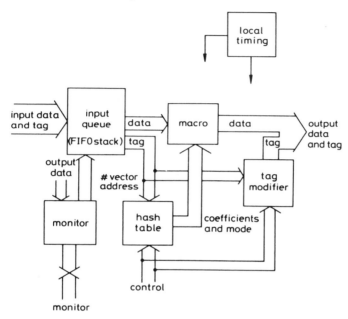

Fig. 22.2 *Generalised primitive structure*

Each data sample or data block gathered by the system is tagged with an identifier as it is sampled which defines the time at which the data was gathered and its source. This tag acts as ticket for the data to ride through the processor network; it is "handed-in" at each primitive node on its route, where it is used to define the process required, is modified and re-applied to the processed data to define the next process. The hash table and tag modifier table performing the control functions are soft loaded from the control system and can be dynamically modified to change the processing schedule.

22.5 Macro development

The arithmetic units which perform the signal processing macro functions can be implemented in variety of ways using current technology. Data flow processors have been developed using networks of single chip digital signal processors (DSP's) (13), micro-coded to define the macro function, multiple high performance array processors (20) and high speed hardware systems, controlled by commercial micro-processors (19). Many first-generation DSP chips (21, 22) that can be used in these types of system are now available from commercial sources. Typically, these contain programmable micro-code ROM store, RAM for data storage and 24 or 32-bit ALU's (both floating point and fixed) and operate at throughputs in the range 3–10 MOPS. These devices are generally designed with internal chip architectures optimised for FFT-type algorithms and are aimed at single processor applications. As a result, it is sometimes difficult to use the first-generation devices in multi-processor systems, and second-generation devices now being developed have the necessary hooks for multi-processor operation (23).

In some applications, for example for IIR filtering the internal architecture of single chip DSP's is not ideal and faster throughput systems require more optimum architecturs using conventional multiplier/accululator based ALU's (24), although this again is likely to change when second generation DSP's become more widely available.

In order to estimate the processing power that could be achieved from DSP devices in the near term (say 3–5 years), it is worthwhile considering the semiconductor technology advances that are likely to occur in this timescale. On past record, commercial pull in the micro/minicomputer field will continue to stimulate development of high-density RAM, ROM and programmable-logic and gate arrays and that the packing densities of these types of devices will continue to double every couple of years.

Current technology DSP's are available fabricated in NMOS (22), bulk CMOS (25), CMOS–SOS (26) and bipolar (24) technologies, with geometries in the range 2–5 microns. With improved process technology over the next few years, device geometries are likely to be reduced, initially to around 2 microns and later to below one. At the same time, it is likely that some current LSI technologies will be dropped as device processing becomes more highly tuned for VLSI (27).

With improved technology, it is likely that devices containing in excess of 40,000 gates could be available with 2 micron geometry and in excess of 120,000 gats with sub-micron devices.

In terms of signal processing complexity, a 16-bit fixed-point real-arithmetic DSP requires around 10,000 gates whilst a 32-bit floating-point complex DSP needs around 40,000. So within the timescale considered, improved technology could provide DSP's for 64-bit floating-point real or 32-bit floating-point (complex) operation. From speed-power considerations, these devices should be capable of operating at throughputs of around 20 MOP's.

Whilst these performance figures represent a major advance over those of current devices, future-generation sonar applications would require 1000 to 10000 such devices to provide the necessry throughput. . . .the integration of such systems should prove an interesting exercise.

22.6 Algorithm development

In view of the potential difficulties involved in integrating distributed, high-throughput systems, it is worthwhile considering ways of reducing overall system complexity. One approach is to speed up the operation of particular macros, particularly those executing functions which represent a major load on the overall signal processing system. Typical of such high load processes are filtering operations for analysis band definition, which represent the major overhead in many passive systems, and Fourier transformation, which often dominates in narrow band detection systems and in pre-processing for adaptive and high resolution processors.

Using conventional FFT algorithms (28), advanced technology transform implementations require around 100,000 gates and potentially operate at transform speeds of around 1-4 milliseconds for a 1024-point complex transform. For sonar systems, with say 5000 processing channels, up to 200 such processors are required to convert data into the frequency domain for further processing.

Various number theoretic methods are available that can be used to some advantage to increase the throughput or reduce the complexity of such processing systems (29). For example, relatively straightforward number theoretic tricks can be applied to the DFT algorithm to produce recursive prime radix transforms (30) with significantly reduced hardware complexity. Further developments of this method have provided greater simplification and bring the hardware required for high precision, wide bandwidth DFTs within the complexity and gate count currently achieved with UHS dipolar, and maybe even GaAs technology within the next few years.

Similar techniques have been applied to constrained coefficient FIR and IIR filter design (31), matrix operators (32), etc, and processing systems using these "alternative low-level" primitives as resources in processor networks are being developed (17).

22.7 Summary

Some of the problems associated with high-throughput digital processing systems for sonar signal processing and the potential impact of advances in semiconductor technology have been outlined. It is anticipated that considerable improvements in device performance will result from the various technology programmes now in place, but that the throughput required for next generation digital signal processing applications will require similar advances in systems architectures and processing algorithms. In these areas, signal processing flow graph architectures, using data-driven processor networks, and low-complexity, high-precision algorithms, making use of simple number-theoretic tricks, are being developed.

Copyright © Controller HMSO, London, 1985.

References

1. See for example: FARAN, J. F. and WELLS, R., 'Correlators for Signal Reception', Harvard Acoustic Lab Tech Memo, No 27,1952. ANDERSON, V. C., 'DELTIC Correlators', Harvard Acoustic Lab Tech Memo, No 37, 1956.
2. ALLEN, W. B. and WESTERFIELD, E. C., 'Correlators and Matched Filters for Sonar', *JASA*, **36**, 1964.
3. See for example: 'Analysis and Design of Integrated Circuits', Motorola Series in Solid-state Electronics, McGraw-Hill, 1967.
4. ANDERSON, V. C., 'Digital Array Phasing', *JASA*, **32**, 1960.
5. REMLEY, W. R., 'Some Effects of Clipping in Array Processing', *JASA*, **39**, 1966.
6. See for example: 'MOSFET in Circuit Design', Texas Instruments Electronics Series, McGraw-Hill, 1967.
7. Proc. of Conf on Real-time, General Purpose, High-Speed Signal Processing Systems for Underwater Research, CP-25, SACLANT ASW Research Center, La Spezia, 1979.
8. See for example: Digests of Technical Papers for recent International, Solid-State Circuits Conferences, ISSCC-82, -83, -84.
9. MORGAN, W. I., 'From VLSI to ULSI', Semiconductor International, May 1984.
10. KAY, S. M. and MARPLE, S. L., 'Spectrum Analysis – A Modern Perspective', *Proc. IEEE*, **69**, No 11, 1981.
11. KNIGHT, W. C. et al., 'Digital Signal Processing for Sonar', *ibid*.
12. MORRIS, L. R., 'A Tale of Two Architectures: TI TMS 320 SPC vs DEC Micro/J-11', Proc. ICASSP 83, Boston, 1983.
13. KNUDSON, M. J., 'MUSEC, a Powerful Network of Signal Processors', Proc. ICASSP 83, Boston, 1983.
14. WU, Y. S., 'A Common Operational Software -ACOS- Approach to a Signal Processing Development System', Proc. ICASSP 83, Boston, 1983.
15. KUNG, H. T., 'Design of Systolic Arrays for System Integration', EUROSIP Int. Conf. on Digital Signal Processing, Florence, 1984.
16. PARKINSON, D., 'High-speed Computing', *Phys. Bull.*, **29**, 1978.
17. CURTIS, T. E., WU, Y. S., CONSTANTINIDES, A. G. and WU, L. J., VLSI Architecture for Signal Processing Alternate Low-Level Primitive Structures -ALPS", Proc ICASSP 84, San Diego, 1984.
18. SWARTZLANDER, E. E. and HEATH, D. J., 'A Routing Algorithm for Signal Processing Networks", *IEEE Trans.* C-28, No 8, 1979.
19. CURTIS, T. E., CONSTANINIDES, A. G. and WICKENDEN, J. T., 'Control Ordered Sonar Hardware – COSH: A Distributed Processor Network for Acoustic Signal Processing', to be published, IEE Proc., part F, 1984.
20. BROWN, N. H., 'The EMSP Data Flow Computer', Proc. of International Conference on System Scienes, 1984.
21. Texas Instruments Inc., TMS 32010 User's Guide, Dallas, 1983.
22. Electronics International, pp. 121–126, May 19, 1983.
23. Advanced data – NEC Image Pipelined Processor, uPD7281, Jan 1984.
24. TRW LSI Products, Multiplier-Accumulator Application Notes, Jan 1980.
25. See for example: Data sheets and application notes no Analogue Devices ADSP family of devices.
26. See for example: Data sheets on Rockwell 31416 multiplier/accumulator Rockwell International Corp., 1983.
27. OLDHAM, H. E. and PARTRIDGE, S. L., 'Comparison of MOS Processes for VLSI', *IEEE Proc.*, **130**, Part I, No. 3, 1983.
28. See for example: 'Digital Processing of Signals' by GOLD, B. and RADER, C. M., McGraw-Hill, 1969.

29. 'Number Theory in Digital Signal Processing', by McCLELLEN, J. H. and RADER, C. M., Prentice-Hall, 1979.
30. CURTIS, T. E. and WICKENDEN, J. T.: 'Hardware-based Fourier Transforms – Algorithms and Architectures', *IEE Proc.*, **130**, Prt F, No. 5, 1983.
31. LIM, Y. C., PARKER, S. R. and CONSTANINIDES, A. G., 'Finite Word-length FIR Filter Design using Integer Programming over a Discrete Coefficient Space', IEEE ASSP-30, 1982.
32. ANG, P. H. and MORF, M., 'Concurrent Array Processor for Fast Eigenvalue Computations', Proc. ICASSP 84, San Diego, 1984.

Chapter 23

Optical fibre systems for signal processing

Prof. B. Culshaw
*Department of Electronic and Electrical Engineering,
University of Strathclyde, 204 George Street, Glasgow, UK.*

23.1 Introduction

A single mode optical fibre is an extremely high bandwidth, very low attenuation transmission medium. For signal processing applications, the large information bandwidth which may be modulated onto an optical carrier implies that time delay-bandwidth products of the order of 10^6 may be realised. This is some two orders of magnitude higher than the TB product which may be achieved using conventional delay line media. Signal bandwidths of perhaps 100 GHz may be contemplated, since the use of the very high (optical) carrier frequency implies very small fractional bandwidth occupancy.

An optical fibre delay line signal processor requires that these advantages be exploited via some suitable technology. The exploitation process may take several forms. At its simplest, the data may be (usually intensity) modulated onto the optical carrier and launched into a conventional fibre optic transmission system. The receiver will then combine predetermined fractions of this data with the input data and retransmit the rest to form a multiple tapped delay line system (see Figure 23.1).

The tapping and combining network is, in this case, entirely electronic and this system is distinguished from an identical electrical network only in the use of the fibre delay line which may offer advantages concerning low radiated signal strengths and compact format. Of more interest is an all-optical delay line signal processor (see Figure 23.2) in which the tapping and recombining are effected in the optical domain. The speed of this processor is not electronically limited, and most of the discussion in this paper will investigate the properties of this class of processor.

All these processors may be most simply realised in a form which deals only with positive algebra. This is somewhat restricted in scope, especially for filtering and related operations, and positive/negative or complex algebra are intrin-

Optical fibre systems for signal processing 203

sically more attractive. The latter requires the use of phase sensitive processing devices in which the phase may be defined either by that of the optical carrier itself, or by a subcarrier (operating for instance at a frequency of a few GHz) in which case, all the processing is done at the subcarrier frequency and the optical medium permits a lower attenuation, lower dispersion and transmission loss than would be obtained in metallic guides. There are, of course, many other contexts into which guided wave optical signal processing is relevant. Fibre optics may be used as the basis of a class of spatial signal processors in which the fibre guides light in an input plane to a predetermined point in an output plane. This may be used to perform simple magnification (as found in, for instance, a simple mobile projection screen), or it may perform more complicated functions such as the discrete Fourier transform. These processors are complicated to assemble and difficult, if not impossible, to program, and so have generated relatively little interest apart from a few very specialised operations.

There are also numerous guided wave approaches to performing analogue signal processing using (usually one-dimensional) versions of normal Fourier optics concepts. These are somewhat outside the scope of this paper, though fibre optics, and perhaps more so integrated optics, will play an important part in their evolution.

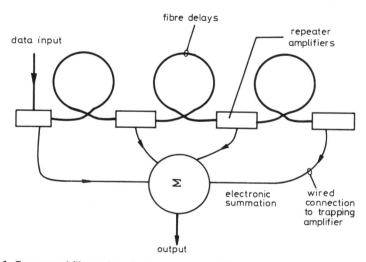

Fig. 23.1 *Transversal filter using electronic taps and fibre optic delay lines*

The remainder of this paper is devoted to a discussion of the general features of fibre optic single mode signal processors with emphasis on technologies in which the processing is implemented entirely in the optical domain. The technology is young, though it is possible to indicate some of the limitations of these concepts and to speculate about future developments in which, perhaps, soliton propagation using a fibre optic amplifier will permit the undistorted transmission of optical pulses over extremely long distances. Perhaps one may then

204 Optical fibre systems for signal processing

conceive of extremely high speed (tens of GHz) systems with delays measured in the hundreds of microseconds. The time bandwidth product of such a system may then exceed 10^7 though the technological developments required to implement these concepts are, of course significant.

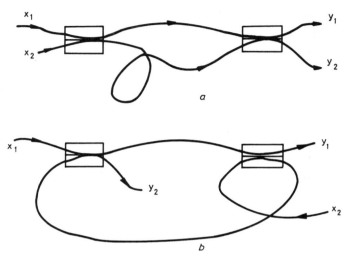

Fig. 23.2 *Fibre optic filters based on optical fibre couplers (a) transversal, (b) recursive*
(after reference 5)

23.2 Components for fibre optic signal processing

The principal required components for a fibre signal processor are shown in Figure 23.2. The data source is modulated onto an optical carrier, indicated as x_1, x_2 in the figure, typically generated using a semi-conducted laser diode. The laser diode must be chosen such that all the recombination delays within the network exceed the coherence length of the laser by a significant margin, and even here, coherent effects will influence the noise level in the system. The modulation must then be applied to the laser diode intensity and positive real taps are the only ones which may be readily realised in optical form, though a hybrid in which negative taps are optically summed and then electronically subtracted from the positive taps may also be demonstrated. If the tap is to be passive, then the only means whereby the signal may be extracted from the line is a fibre optic directional coupler. It should be emphasized that unless gain can be built into the system this signal extraction process will, in itself, dominate the attenuation of the delay medium and, therefore, limit the TB product. The delay lines may obviously be configured as transversal or recursive processors as indicated in the Figure.

The key technology here is then the delay line tap. Many forms of fibre optic coupler have been tried though the only one which seems to be suitable for laboratory assessments of signal processor concepts is the block coupler originally proposed by the team at Stanford University. This device can be made with an extremely low loss and the adjustable coupling coefficient makes it readily suitable for investigations of numerous processor formats. To date, this device has not been realised in an environmentally rugged form and this could prove to be a major disadvantage. This particular tap may, of course, be incorporated into an interferometric signal processor which would use coherent techniques. However, accurate long term control of the optical phase in the fibres between the taps would probably prove to be difficult in all but the simplest configurations. It is perhaps in the area of extremely high speed signal processing, maybe 100 GHz and above, that a coherent system could be feasible, but based on integrated optic implementations of the concepts.

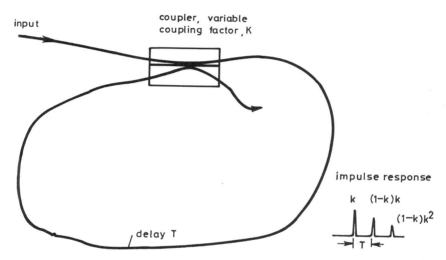

Fig. 23.3 *Fibre optic recursive filter using single coupler*

23.3 Some examples of fibre optic signal processing

The simplest single coupler processor which may be configured is the transversal filter shown in Figure 23.3. The impulse response for this system is also shown in this Figure. If we take the special case where K, the coupling coefficient, is taken to be 0.83, then we get a frequency response, $F(\omega)$, which is very close to $\cos(\omega T/2)$ where T is the delay around the loop. Laboratory demonstrations of such a filter have confirmed notch depths exceeding 50 dB up to frequencies exceeding 1 GHz. Again, the low dispersion, low attenuation properties of the medium are essential to achieving this high performance. If we make K approximately unity, then we have a recursive filter in which the impulse response is a large single first pulse followed by a succession of smaller pulses tapering off by an amount determined by the loop attenuation and the loss in the coupler. This impulse response gives rise to an equivalent comb-shaped frequency response and a pass band filter can therefore be realised. Again, this has been experimentally demonstrated, though the requirement for a very high value of K implies that relatively little optical power remains in the system to perform the filtering function of interest (defined by all the pulses in the impulse response apart from the first) so that there are significant signal to noise ratio problems with this type of device. Incidentally, the form of the filter in Figure 23.3 with K approaching unity, is, in optical terms, equivalent to a Fabry-Perot filter. However, since the illuminating source is not coherent, this aspect of the performance of the device may be initially neglected.

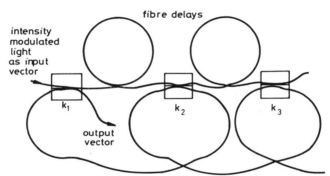

Fig. 23.4 *Fibre optic matrix-vector multiplier coupling coefficients k_1, k_2 and k_3 define the matrix elements*

In common with other delay line technologies, any function which requires the addition of delayed samples of one or a number of signals to be added and/or multiplied together in some predetermined fashion can be realised in fibre optics. Among these functions we may include the fibre optic matrix-vector multiplier shown in Figure 23.4. This concept may be extended from the 2×2 case illustrated to any $N \times N$ Teoplitz matrix multiplication operation. There is clearly considerable scope for further applications of these ideas and many of the emerging concepts involving pipeline processors and similar ideas may be readily applied here.

23.4 Some basic limitations on all optical monomode fibre processors

Shot noise is the feature which determines the absolute best performance achievable in an optical system. If we apply this to be the case of a fibre optic tapped delay line and we assume that we are processing analogue signals to a predetermined signal to noise ratio, then the shot noise to signal ratio at each tap should reach the specified minimum level. Clearly, the exact power requirements depend upon the specification, but we may analyse the case of a 40 dB signal to noise ratio in a delay line in which there are 50 taps and for which the signal bandwith is 1 GHz. Asuming that the noise contributions from the various taps are uncorrelated, this gives a total required launched optical power into the system of 0.2 W at a wavelength of 850 nm. This indicates one of the principal limitations of very high speed analogue signal processors, since any reasonable signal to noise ratio will involve very high optical powers. There is, of course, the square root dependence of signal to noise ratio on optical power, which means that halving the required SNR to 20 dB reduces this launched optical power requirement by 40 dB. There is clearly an even greater than normal incentive to operate in the digital domain at very high bit rates in order to process the same amount of information. Thus, perhaps a very high speed analogue to digital convertor operating with an accuracy of maybe 7 bits would be an extremely useful input device to an optical signal processor. On the other hand, if we are to exploit the potential offered by, for instance, soliton propagation and/or continuous amplification effects in the fibre, then these relatively high power levels are necessary in order to access the required basic physics.

This very simple calculation of the power required in the processor for a given bandwidth and signal to noise ratio has very rapidly demonstrated that any high speed all optical signal processor will, in all probability, be digital. However, current technology, and that available for the forseeable future, implies that analogue data will be all that is accessible. The question of the stability of the analogue tap is, then, extremely important, and in the devices demonstrated to date temperature coefficients of the order of fraction of a percent per degree are typical. Thus, very high precision processing would, of necessity, involve precise temperature stabilisation. In this particularly simple form of the processor, the taps are limited to real quantities and a coherent processor would be attractive, though again if we extrapolate to a digital system, this requirement will be relaxed to, at most, positive and negative taps. There are also other technological constraints on the operation of these processors and these include achievable detector speeds, achievable modulator speeds and accuracies, and extraneous noise sources. Of these, the most serious appear to be the effects of coherent recombination of phase noise in the laser signal used to energise the system, and intensity noise phenomena caused by relaxation effects, etc. and present in the output spectrum in the source. These contributions may be avoided by using an incoherent source, though it is unlikely that sufficiently high output powers can be launched into single mode fibres from such sources.

23.5 Future trends and possibilities

The very high speed and relatively long delays which can be obtained with conventional optical fibre used in a signal processing configuration are but a part of the overall potential of this technology. Possible future systems would include a totally coherent processor in which complex functions may be realised, though it is likely that this will be a short device based on integrated optics rather than fibres. There are other important considerations which must be addressed before a practical system can be realised. Of these, the most obvious is a requirement for some optical domain preprocessing of the input and output signals prior to further electronic processing. This is required simply because the speed of the optical system exceeds that of any known or projected electronic equivalent.

Finally, it is interesting to extrapolate and guess what the technology may be capable of in the future. Fibre optic Raman amplifiers have been demonstrated and this phenomenon could be used to compensate for attenuation and coupling losses in a fibre optic tapped delay line. pulsed dispersion can be made identically zero by using solitons, so that, in principle, digits of duration a few picoseconds can be sent for hundreds of kilometres with negligible attenuation or dispersion. TB products approaching 10^8 could be envisaged with such a system, though, of course, the optical power levels required are high and such a processor would be bulky and exhibit a relatively high power dissipation, simply because of the requirement for high power optical sources. It is difficult to predict whether the eventual realisation of such a processor is either necessary or desirable, but without doubt the prospect that such a powerful system may exist is undeniably intriguing.

23.6 References

1. International Optical Computing Conferences, Proceedings published by IEEE, most recent Cambridge, Mass., 1983.
2. B. CULSHAW, 'Fibre optic sensing and signal processing', Peter Peregrinus, 1984.
3. Proceedings CLEO (Conference on Lasers and Electro-optics) published annually by Optical Society of America.
4. J. W. GARDNER, 'Introduction to Fourier Optics', McGraw Hill, 1968.
5. Most of the concepts described in this paper are based on Stanford University work in this area. Proc. IEEE July 1984 contains a more detailed account of this and of other related work in optical signal processing.

Chapter 24
Satellite communications
J. L. Everett
R.S.R.E., Defford, UK

24.1 Background

The continuing high demand for global, regional and domestic communications is being satisfied by a variety of satellite systems of ever increasing capability and complexity. It was not until 1963 that rockets sufficiently powerful to launch satellites into geostationary orbit were available, and SYNCOM II was the first to lie on the geostationary arc 35,800 km above the equator.

INTELSAT (the international telecommunications satellite consortium) commenced operations in the mid 60's. INTELSAT provides multichannel trunking facilities for international carriers as well as leasing transponders to individual nations for regional or domestic systems. WESTAR, RCA, SATCOM, Satellite Business Systems (SBS) and EUTELSAT are examples of domestic/regional systems. INMARSAT provides ship/shore communications in the Atlantic, Pacific and Indian Ocean regions, and a specialized search and rescue system exists in the form of SARSAT. High-quality signals are transmitted over satellites to cable network link-ends, and more recently, directly to consumers in their homes. Direct-broadcasting should arrive in the U.K. after the launch of UNISAT, and should that system not materialize, it will be one of the services available from OLYMPUS.

24.2 Satellite system operation

The dense population of satellites on the geostationary arc requires international agreement to prevent interference with terrestrial as well as other satellite communications links. Up-link and down-link frequency bands and angular spacing on the geostationary arc are determined by these agreements, and Fig. 24.1 depicts current occupancy of the arc. Spacing on the geostationary arc cannot be too close otherwise an up-link to one satellite will also be received by adjacent satellites, causing interference. Strict constraints are placed upon aerial-beam profiles to ensure these "crosstalk" effects are kept under control.

210 Satellite communications

C-band satellites are placed 4° apart on the geostationary arc and all of the popular slots (those serving the Atlantic area and the North American continent) are now occupied. The use of higher frequencies allows closer spacing on the arc, and this, together with the wider bandwidths allocated at these frequencies, provides extra capacity. Communications at 14/11 GHz are now well established and 30/20 GHz systems are appearing. The path losses at higher frequencies increase significantly in the presence of precipitation and this phenomenon is the subject of extensive investigation: clearly, sufficient operating margins must be included in the link budget to accommodate these effects.

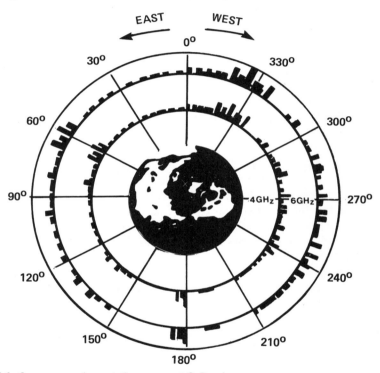

Fig. 24.1 *Occupancy of geostationary arc at C-Band*

24.3 System studies and link budgets

The basic up-link equation for a single radio frequency carrier has been derived by Feher (1):

$$\left(\frac{C_u}{N_{ou}}\right)_{dB} = \underbrace{10 \log P_T G_T}_{\text{earth station e.i.r.p.}} - \underbrace{20 \log \frac{4\pi R_u}{\lambda_u}}_{\substack{\text{"free space"} \\ \text{up-link loss}}} + \underbrace{10 \log \frac{G_{su}}{T_s}}_{\substack{\text{satellite} \\ \text{G/T}}} + \underbrace{10 \log L_u}_{\substack{\text{additional} \\ \text{up-link losses}}} - 10 \log k$$

(24.1)

and the corresponding down-link equation is given by:

$$\left(\frac{C_d}{N_{od}}\right)_{dB} = \underbrace{10\log(P_sG_{sd})}_{\substack{\text{satellite}\\\text{e.i.r.p.}}} - \underbrace{20\log\frac{4\pi R_d}{\lambda_d}}_{\substack{\text{"free-space"}\\\text{down-link}\\\text{loss}}} + \underbrace{10\log\frac{G_d}{T_d}}_{\substack{\text{earth station}\\\text{G/T}}} + \underbrace{10\log L_d}_{\substack{\text{additional}\\\text{down-link}\\\text{losses}}} - 10\log k$$

(24.2)

where P is transmitted power, G is antenna gain, R is path length, λ is wavelength, T is effective noise temperature, k is Boltzmann's constant ($k = 1.38 \times 10^{-23}$ J/Kelvin) and L_A is the additional loss in the path. L_A includes loss caused by precipitation, losses in the transmission system between power amplifier and antenna, antenna pointing error loss and receiving system loss on the satellite. Equations (24.1) and (24.2) are often written in the simplified form:

$$\left(\frac{C}{N}\right)_{dB} = \text{e.i.r.p.} + \frac{G}{T} - L - L_A - k \quad (24.3)$$

In a conventional frequency translation satellite system, the overall carrier power-to-noise density ratio $(C/N_0)_T$ at the receiving earth station is:

$$\left(\frac{C}{N_0}\right)_T = \frac{1}{(N_{ou}/C_u) + (N_{od}/C_d)} \quad (24.4)$$

The link budget equations are a critical part of the satellite communications system design process since they relate power levels and other parameters to a specified system performance. Down-link budgets are controlled largely by the r.f. power available on the satellite, which in turn is limited by the electrical power available from the solar cells and the efficiency of the r.f. source (travelling-wave tube or solid state device). The general form of up-link and down-link signal power budgets has been represented diagrammatically by Cuccia (2), Fig. 24.2.

24.4 OLYMPUS

Reference has already been made to several different satellite systems, but one which embodies a number of advanced technical features and serves as a useful illustration of the next generation of communications satellites is the OLYMPUS family, of which the first is OLYMPUS-1.

OLYMPUS-1 is the name assigned to the Large Telecommunication Satellite, previously known as L-SAT, currently being built for the European Space Agency by an international consortium led by British Aerospace. This is an experimental 3-axis stabilized satellite with an intended mission lifetime of 7 years. Launch options include various versions of ARIANE and the Space

Transportation System (Space Shuttle) and OLYMPUS-1 will be placed into a geostationary orbit of longitude 19°W in 1987.

The design of the OLYMPUS family has been strongly influenced by anticipated future communications requirements based on extensive market surveys and an awareness of competing systems.

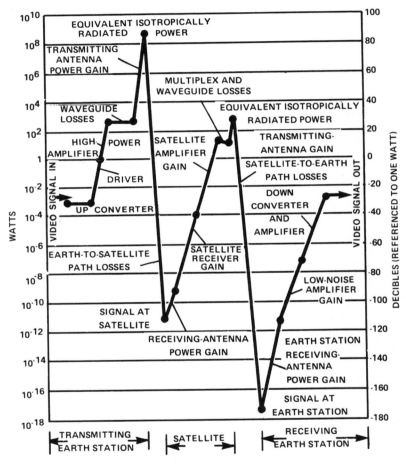

Fig. 24.2 *Link signal power budget diagram*

Four payloads will be flown on OLYMPUS-1: these are:-

Payload 1: Beacons operating at 12/20/30 GHz.
Payload 2: A Specialised Services Payload operating at 14/12 GHz.
Payload 3: A TV Broadcast Payload.
Payload 4: A 30/20 GHz Communications Payload.

These payloads are now detailed with the advanced features highlighted.

24.4.1 *Payload 1*

Propagation data is required for the prediction of the availability of satellite links operating at particular frequencies. This requires the formulation of realistic and reliable models for signal loss, depolarization and other features which may degrade the propagation of radio waves between earth and the satellite. This data is required in particular at frequencies of 30 and 20 GHz which have been allocated for satellite communications, and the beacons on OLYMPUS-1 have been included for this purpose. The beacon sources have been designed to provide signals with high stability in amplitude and frequency, polarization purity and minimal phase jitter.

24.4.2 *Payload 2*

The Specialized Services Payload is included for the provision of pilot business services utilizing small earth terminals with antenna diameters of less than 3 m, for the demonstration of multipoint video-teleconference by satellite and television programme distribution to cable networks. This payload will operate with up-links in the 14 or 13 GHz bands and a down-link at 12 GHz. A steerable multibeam antenna is proposed with 5 separate spot-beams each just over 1° diameter, with frequency reuse between them. The satellite effective isotropic radiated power (e.i.r.p.) will be about 45 dBW for each spot-beam. A 4 × 4 switching matrix will be integrated into the payload to allow the transmission of satellite-switched time-division-multiple-access (s.s.-t.d.m.a.) signal formats. The basic principle t.d.m.a. is described in the next section: the satellite-switched capability means that individual signal bursts can be re-routed by the satellite in a dynamic manner.

24.4.3 *Payload 3*

A 2-channel payload suitable for direct broadcasting will be carried: this will offer satellite-to-home broadcasting over a small geographical area or satellite-to-cable head installations on a pan-European basis. One of the channels will be assigned to pre-operational trials in Italy with the other shared by the remainder of the European participants in the project. The receive antenna on the satellite will accept a 17 GHz up-link from anywhere within Europe and will be compatible with up-links from outside broadcast stations. Signals will be transmitted from the satellite in the 11.7 to 12.5 GHz band. An output e.i.r.p. of about 60 dBW will be available from each 27 MHz wide channel.

24.4.4 *Payload 4*

The 30/20 GHz communications package has been configured to support a range of experiments and demonstrations. Independently steerable receive and transmit antennas on the satellite each generate a beam 1.2° in diameter which can be used for point-to-point video teleconference or data transmission. The exceptionally wide-bandwidth transponders (250-MHz wide) offer the prospect of some important and exciting experiments in wide-band communications.

24.5 Advanced modulation techniques

Early satellite communications systems utilized frequency modulation (f.m.) schemes where the information signals are superimposed on a carrier and then transmitted. There is now a strong trend towards digital transmission systems, principally phase modulation formats based on Phase Shift Keying (p.s.k.) or Quadrature Phase Shift Keying (q.p.s.k.) in which phase transitions of the carrier can occur at every 180° or 90° respectively representing the digital information being transmitted over the link. Digital modulation is more resistant than analogue modulation to errors in non-linear transponder channels. The relationship between error rate and carrier-to-noise ratio for several different types of digital modulation is shown in Fig. 24.3.

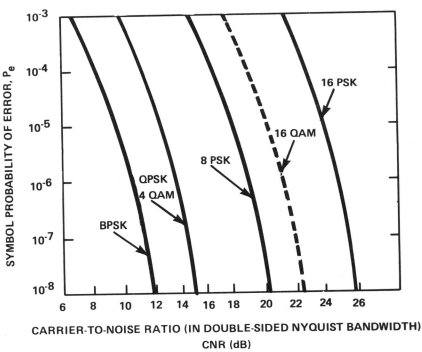

Fig. 24.3 *Relationship between error rate and carrier-to-noise ratio*

Two different types of digital modulation system used in modern satellite communications are described below:

24.5.1 *Time-division multiple access (t.d.m.a.)*

Fig. 24.4 illustrates the basic principle of t.d.m.a. A number of ground stations transmit burst signals (each containing many bits of information) to a satellite so that the bursts arrive in a sequential manner without collision: these bursts are down-converted in the satellite and appear on the down-link received by the earth terminal. Two major advantages are obtained from using t.d.m.a.: each burst occupies the full transponder so that intermodulation products (caused by the transponders being operated in a near-limiting condition) will not cause interference with other signals, and digital signalling formats on the satellite link are compatible with the general trend towards digital networks on the ground.

s.s.-t.d.m.a. allows dynamic switching on-board the satellite, giving the satellite the functional appearance of a "switchboard in the sky". This will permit the combined terrestrial and space segments to be utilized in a far more efficient manner than in conventional systems.

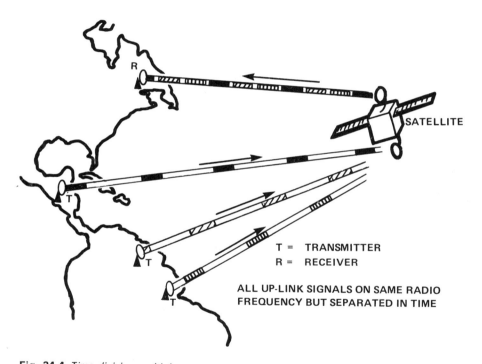

Fig. 24.4 *Time division multiple access*

24.5.2 *Spread spectrum*

Spread spectrum is a method of transmission where the signal occupies a bandwidth in excess of that necessary to send the information: the band spread is accomplished by mixing the signal with a high bit-rate code which is independent of the data, and synchronised reception with a locally generated replica of the code at the receiver is used for de-spreading and subsequent data recovery.

This description specifically excludes modulation formats such as f.m., where spreading is a function of the information being transmitted.

The concept of artificially increasing the bandwidth of the transmitted signal may appear strange, but spread-spectrum modulation formats offer a number of substantial advantages to a communicator: low spectral energy density (minimizes interference with other users), low probability of intercept (unobtrusive communications), resistance to jamming, and multiple access (several communicators can share the same channel). There are two fundamental types of spread spectrum, direct sequence (or psuedo-noise) and frequency hopping: the former type is normally used in satellite communications. This is a form of modulation which has obvious military applications.

24.6 The future

The continuing requirement for more bandwidth and more comprehensive services will continue to provide the motivation for innovative technical developments in satellite communications. The quest for greater bandwidth will be satisfied by the higher frequency bands with their larger bandwidths, by spatial reuse of frequency achieved from the use of zone and spot beams, and by polarization diversity. Improved efficiency modulation formats will be introduced, and modulation and coding schemes will coalesce in an increasingly digital environment where t.d.m.a. and demand assignment packet transmissions will become commonplace. Intelligent switching under computer control will allow rapid reconfiguration of satellite transponders, and data management on board the satellite together with error detection and correction will occur in a manner transparent to the system user.

A range of business services and t.v. receive-only systems will increase their share of satellite transponders and the expansion in the small-terminal market to support these communications in the next decade will demand low-cost receivers with integrated front ends comprising the antenna, feed, low-noise downconverter and demodulator: these will use digital modulation formats, including t.d.m.a. and also direct-sequence spread spectrum for some specialist applications.

Satellite-to-satellite communications will utilize millimetre wave and even optical laser links to support data rates in the gigabit per second range.

Small terminals will shrink in size: manpack systems are already in use by the military and by the search and rescue services, and hand held paging systems are being designed. A satellite link to augment cellular mobile radio-communication systems has been proposed for use in the U.S.A.

In conclusion, the exciting and innovative development in satellite communications over the past two decades will be continued with even more adventurous systems and concepts in the future.

24.7 References

1. FEHER, K., 1981, Digital Communications (Satellite/Earth Station engineering), Prentice-Hall Inc.
2. C. LOUIS CUCCIA., (Editor), 1979, The Handbook of Digital Communications, E. W. Communications Inc.

Copyright © Controller, HMSO, London 1985

Chapter 25
A VLSI array processor for image and signal processing

Stephen Pass
Marconi Command and Control Systems Limited, Frimley, UK

25.1 Introduction

The GEC Rectangular Image and Data ("GRID" – see footnote) processor has the typical SIMD architecture exhibited by such machines as the UCL CLIP4 (Duff, 1980), ICL DAP (Reddaway, 1973) and NASA MPP (Batcher 1980). It has been developed to provide a powerful and flexible system suitable for image and signal processing applications. The major system components have been designed in custom VLSI, giving the GRID a considerable size advantage over existing parallel array processors.

This paper describes both the hardware and software aspects of the GRID system and presents some of the algorithm work relating to the system's image and signal processing capabilities.

25.2 System architecture

The GRID computer consists of an interconnected array of simple, bit-serial processing elements (PEs) arranged on a square mesh, each with its own local memory. A central controller broadcasts a sequence of instructions to the PEs which then operate in an identical manner on their own data.

Each PE contains an ALU, 4 special purpose registers, 64 general purpose registers, connections to the eight neighbouring PEs and a port to external memory. A custom VLSI circuit containing 32 PEs has been designed in 3 μm bulk CMOS (with double-layer metal). A 64 × 64 PE array with 8 K bits of external memory per PE will be constructed for the prototype GRID system.

The central controller is divided into two principal blocks. The first is known as the "array controller" since it deals with the low level interface to the PE

"GRID" is an integral project name and has no connection with Grid Computer Systems Ltd.

A VLSI array processor for image and signal processing

array. It is a microprogrammed unit and supports the primitive array operations which include arithmetic, comparison and routing (i.e. moving data across the array). Note that since each PE operates in a bit-serial manner even the most basic operations (e.g. addition) have to be performed under microprogram control.

The array controller also provides hardware support for mapping large data sets (e.g. 512 × 512 images) onto the PE array. This ties in with the high-level notion of unconstrained arrays which will be covered in the next section. Three different custom VLSI circuits implement the specialised function of the array controller.

The second block is known as the "scalar processor" since it handles the serial aspect of task execution. These include both serial control flow and scalar arithmetic, i.e. arithmetic operations with a very low degree of parallelism where it would be inappropriate to use the PE array. An MC68000 microprocessor has been selected for this role in the prototype GRID system.

25.3 Programming

The GRID system will be connected to a host computer which provides a multi-user environment for the development and maintenance of both software and documentation. Most user programming will be performed in a parallel high-level language which is being developed from 'C' (Kernighan and Ritchie, 1978). The parallel extensions to 'C' include constructs for both data declaration and conditional execution. Parallel data declaration is interesting in that it is unconstrained by the size of the PE array. The user can specify an array of the required size and the underlying system hardware and software maps it invisibly onto the PE array.

For applications where performance is crucial it will be possible to program in array controller assembler, or even microcode. It is anticipated that frequently used functions (e.g. kernel FFT operations, convolution and median filters) will be programmed at this level.

25.4 Applications

A variety of image and radar signal processing application have been examined with the aid of a software simulator. A number of examples, together with the performance expected from a GRID system are listed below:

Using a 64 × 64 PE array

Arithmetic operations on a 64 × 64 data array

    8-bit addition $2.5\,\mu s$ = 1.6 GOPS
    8-bit multiplication $10.3\,\mu s$ = 0.4 GOPS

220 A VLSI array processor for image and signal processing

Image processing operations on a 256 × 256 × 8-bit image

3 × 3 convolution (arbitrary 8-bit weights)	1.1 ms	= 60 MOPS
3 × 3 median	2.0 ms	= 33 MOPS

4096-point complex FFTs (16 I + 16 Q bits)

2 FFTs computed simultaneously	1.1 ms
16 FFTs computed simultaneously	8.5 ms

25.5 References

BATCHER, K. E., Architecture of a Massively Parallel Processor, Proc. 7th Annual Sym. on Computer Architecture, 1980, pp. 168–173.

DUFF, M. J. B., Parallel Processors for Digital Image processing. Advances in Digital Image Processing, Plenum, 1980, pp. 256–276.

KERNIGHAN, B. W., RITCHIE, R. M., The C Programming Language. Prentice-Hall, 1978.

REDDAWAY, S. F., DAP – A Distributed Array processor. Proc. 1st Annual Sym. on Computer Architecture, 1973, pp. 61–65.

Chapter 26
VHPIC for radar
C. Pell, A. C. Fairhead and M. B. Thomas
Royal Signals and Radar Establishment, Malvern, UK

26.1 Introduction

This paper discusses the future signal processing technology required for radar systems. Although signal processing can be considered to begin on the radio frequencies right at the antenna array, we shall confine our attention to post-r.f. processing, commencing with digital processing and beamforming for the antenna array and following the data stream to the fully-processed plot or track outputs. We shall use the phrase Very-High-Performance Integrated Circuits (VHPIC) without reference to any particular technology programme. Similarly, we shall not refer to particular applications of radar, because we consider that all radars, whether on the battlefield, airborne, or for long-range ground-based air defence, will tend towards multi-mode operation, requiring flexible, coherent processing, adaptive antenna processing and false-alarm control.

26.2 Radar processing functions

The relative importance of performance attributes does vary with application. For example, a future airborne radar for air defence, performing search (range and velocity), track, air-to-air missile illumination, Doppler beam-sharpening, surface search, and ground mapping would require a high degree of flexibility, compactness and low power consumption. A future bistatic radar for long-range, ground-based air defence, performing volumetric search (with and without ecm), limited-sector ecm analysis, passive detection and analysis, and multiple waveform reception would again require flexibility but not necessarily compactness or low power consumption. Reconfigurability and sheer processing capacity would be of paramount importance. A generic radar system is depicted in Fig. 26.1. The key functions are considered to be array antenna processing, programmable matched filtering (pulse compression), coherent and incoherent processing (integration), high-resolution processing, and plot extraction (ambiguity resolution, constant false-alarm-rate (CFAR) processing, etc).

222 VHPIC for radar

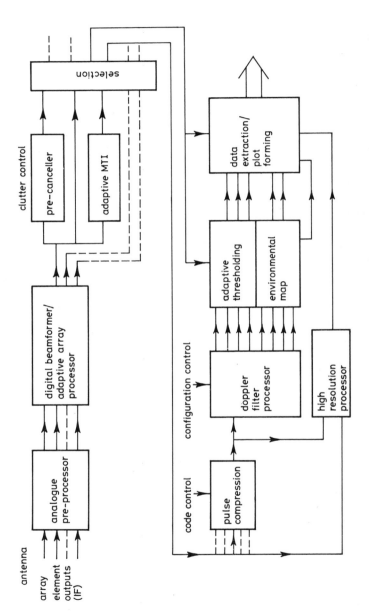

Fig. 26.1 *Radar Signal Processing Architecture*

These five functions will now be described in order to introduce the algorithms, processes and performance requirements.

26.2.1 Array antenna processing

The most obvious process is digital beamforming (1). Digitised complex element- or subarray-signals are multiplied by complex digital weights, which define the beam steering angle and sidelobe level, and summed to a single beam with quadrature components, I_s and Q_s

$$I_s = \sum_{i=1}^{N} (I_i \cdot I_{wi} - Q_i \cdot Q_{wi})$$

$$Q_s = \sum_{i=1}^{N} (I_i \cdot Q_{wi} + Q_i \cdot I_{wi})$$

where N is the number of elements, I_i and Q_i are the ith element signals, and I_{wi} and Q_{wi} are the ith element weights. The full advantage of digital beamforming is only obtained when the weights are dynamically updated to account for system errors.

The beam pattern may be dynamically adapted to overcome jamming or severe clutter using interference-only or total-power minimisation (2). Adaptive algorithms are exemplified by Widrow's least-mean-squares technique (3). Where fast adaptivity is important, direct sample matrix inversion (SMI) (4) is attractive although computationally intensive. Hung and Turner (5) have proposed a less intensive direct technique based on vector operations.

26.2.2 Programmable pulse compression

Because of their limited peak power of transmission, modern solid-state radars, in particular, transmit the energy required for long-range detection as a long, phase- or frequency-modulated pulse. In the receiver, the pulse's non-linear phase spectrum is realigned by a matched filter, so that its component frequencies construct a short pulse, giving the required detectability and range resolution. Additional spectrum weighting is used to reduce "range sidelobes" (6).

A matched filter has a finite impulse response (f.i.r.), defined as the complex conjugate of the waveform it is matched to, reversed in time. A non-recursive digital filter provides a direct way of convolving this with the input signal. It must store enough coefficients to satisfy the Nyquist sampling criterion, and therefore its complexity is proportional to the time-bandwidth product of the waveform (7). For high time-bandwidth product waveforms, less hardware is required for indirect convolution (8), which involves a discrete Fourier transform (DFT) of the input signal, multiplication of the DFT of the filter's impulse response, and an inverse DFT to obtain the output waveform. Nevertheless, matched filters (or those deliberately mismatched by weighting) have traditionally been analogue, a digital implementation having been too bulky and power consuming, or not fast enough to operate in real-time, except for a few applications (9, 10). VHPIC will be a timely remedy, because the trend towards multi-mode radars, which transmit different waveforms for different purposes, calls for a programmable matched filter. This not only enables the radar to operate in different modes, but also allows the use of waveform- or code-agility as a useful feature in electronic warfare. In addition, it can be made to maintain the optimum gain in peak power for the signal being compressed over other, non-white, types of background noise, whatever the source of that noise (7). Another possibility is compensation for amplitude and phase errors introduced by other hardware in the system, which otherwise would create unacceptably high range sidelobes around the compressed pulse (11).

26.2.3 Coherent and incoherent processing

Spectral processing allows moving targets to be discriminated from stationary and/or moving clutter. Ground-based radars have until now formed only a single "signal-pass" filter, with Moving-Target Indicator (MTI) or Adaptive MTI systems. Modern airborne and future ground-based radars require multiple filters. A bank of up to eight classical f.i.r. filters is feasible, and allows each channel to be separately optimised. For larger numbers (32 is a common requirement), FFT techniques are preferred.

In an incoherent processor, the function prior to decision estimation is simple integration, requiring single- or multi-bit stores and summation.

26.2.4 High resolution processing

Future radars are likely to require high-resolution analysis of targets or the environment, which can be performed in the spatial or spectral domain. A high-bandwidth waveform will give high range-resolution for spatial analysis. Spectral analysis can be performed using conventional long-integration-time Fourier techniques, but in the future, algorithms such as maximum entropy and maximum likelihood might become practicable real-time processes.

26.2.5 Plot extraction

Radar users are tending to require displays of plot-extracted data rather than "raw" data. Plot-extraction includes CFAR processing, peak detection, ambiguity resolution, parameter estimation and refinement, coordinate conversion, plot message assembly.

At least one stage of CFAR is likely to include thresholding based upon background estimation. This adaptive process may use one- or two-dimensional cell-averaging, although non-parametric processes such as rank-ordered detection are attractive. The complexity of the many detectors/thresholders associated with a Doppler filter bank becomes significant.

Ambiguity resolution is correlation performed on parametric selections of data from a target observation. Often, an unfolding process in range and velocity is followed by coincidence detection.

26.3 Processes and performance requirements

This section estimates the arithmetic requirements of the processes described in section 1, taking array antenna processing, programmable pulse compression and coherent Doppler processing as examples.

26.3.1 Array antenna processing

The basis of digital beamforming is complex multiplication (requiring four real multiplications and two additions) and accumulation. Its complexity depends on the number of array elements and the number of beams to be simultaneously formed, whilst the speed is dictated by the waveform bandwidth. The requirements are summarised below:

Number of elements (K)	Number of beams	Waveform bandwidth (MHz)	Arithmetic operations per second	Number of multipliers required
100	5	2	4×10^9	320
100	10	10	4×10^{10}	3,200
100	50	10	2×10^{11}	16,000
50	2	5	2×10^9	160

We define one arithmetic operation as $A*B + C \to C$ and estimate the number of multipliers required assuming an 80 ns execution time which is typical of fastest devices presently available (12-bit or 16-bit). The first 3 examples apply to multibeam bistatic radar receivers whilst the last applies to a dual-elevation-beam monostatic system. In all cases, an advanced VHPIC technology is required to allow a practical implementation. More efficient algorithms such as the FFT can be used but individual beam optimisation is then precluded.

Where dynamic pattern adaptivity is performed for unwanted signal cancellation, the number of multiply and add operations required is high, being proportional to K^3 in the case of the SMI technique. Hung and Turner's algorithm (5) reduces the number of arithmetic operations required to $(4M^2 + 6M + 2)K$ real adds and $(4M^2 + 8M + 4)K$ real multiplies. Here M is the number of noise samples required and is approximately equal to the number of signals to be suppressed.

26.3.2 Digital pulse compression

The direct non-recursive filter with an impulse response N-samples long must perform N complex multiplications for every output sample it produces. The alternative indirect method using radix-2 FFTs requires $4 + 2\log_2 N'$, where N' is the smallest integer power of 2 greater than N. As N increases, the indirect method becomes more efficient in terms of complex multiplications at $N = 13$, because it then requires only $4 + 2\log_2 16 = 12$. However, if an indirect algorithm that required fewer multiplications were used, such as Winograd's (12), the cross-over point might be lower. Alternatively, various efficient direct-convolution algorithms (13) would raise it in favour of the direct method. One architecture for direct convolution which is particularly attractive because of its regularity and simplicity is the systolic array (14). This shows that memory and control, not just arithmetic, should be considered when evaluating algorithms and architectures. The proposal of a "network cost" (15) is a step towards taking these factors into account.

Nevertheless, a programmable matched filter would probably have to process waveforms with a time-bandwidth product up to 1,000 (for example, a long-range search waveform with a duration of $100\,\mu s$ and a bandwidth of 10 MHz), and the indirect method is likely to be the more efficient. 2048-point FFTs could process sequences of length 1024 samples (10). Since it is difficult to predict the details of the architecture, we shall determine the performance required only in terms of the butterfly, not memory or control. Architectures are judged in more detail in the literature (15, 16, 17), the last reference showing how the architecture may be optimised for a particular technology. If radix-2 FFTs were used, each would need to perform $1 + \log_2 N'$, or 11, butterflies per output sample. The 10-MHz sample rate of the long-range search waveform therefore requires, for each transform, 440 M real multiplications/s, achievable at present with about 36 devices. Any waveform with a time-bandwidth product less than 1024 and a bandwidth lower than 10 MHz may be processed by the same filter. 16-bit internal arithmetic appears to be ample (9).

26.3.3 *Doppler processing*

A small number of Doppler filters may be synthesised with conventional f.i.r. filters. Delay/store, multiply and add operations are required in addition to weights computation.

In an FFT-based Doppler processor, between 8 and 64 filters are commonly needed for suveillance, and larger numbers for detailed spectral analysis. The requirements for a 32-channel Doppler processor are summarised below:

Number of beams	Waveform bandwidth (MHz)	MOPS	Number of multipliers required
1	2	20	2
5	10	5×10^2	50
50	10	5×10^3	500

Although less demanding than digital beamforming, a VHPIC-based processor is dictated. Additionally the log-modulus extraction ($\log I^2 + Q^2$) and adaptive CFAR which nomally follow each filter channel, incorporate high-speed multiplication, addition, storage and special functions such as successive approximation.

Of the three processes we have chosen as examples, beam-forming is the most demanding in terms of complex multiplication, which is the highest-level arithmetic operation common to all three; then pulse compression and then Doppler processing.

26.4 VHPIC technology

26.3.1 *Introduction*

The processes identified in section 2 can be loosely classed as inter-domain, for example DFT and FFT; and intra-domain, for example arithmetic operation, modulus extraction, correlation, encoding, direct convolution and signal conditioning (such as CFAR). The former often demand parallel architectures which are attractive for VHPIC because of their regular structure and short interconnections. The latter are feasible as complete specialised functions on one chip, taking advantage of VHPIC's high packing density to incorporate and connect many different sub-functions. This allows higher speed, lower power consumption and ease of applicability. Although the technology chosen for a function will depend on speed, power dissipation and packing density, of increasing importance are the economics of manufacture (the number of masks required and the difficulty of attaining them). Note also that as lines are reduced to sub-micron dimensions, some form of on-chip supply reduction to 3 V (say) is necessary to reduce voltage stress across them (18).

26.4.2 *Building blocks*

26.4.2.1 *Analogue-to-digital converter (ADC)*

Analogue radar signals must be digitised at a rate commensurate with their bandwidth. An example of an ADC in current use is Analogue Devices, which has 12-bit resolution and 5 MHz sampling rate. It contains hybrid microcircuits, ICs and discrete components on a 7" × 5" board, and consumes 13 W. The future requirement is for a cheaper single-chip ADC, giving 16 bits at 10 MHz or faster, and minimum power consumption.

26.4.2.2 *Complex multiplier (19)*

A complex multiplier is currently being developed with MOD DCVD support at the GEC Hirst Research Centre in 3 μm CMOS-SOS technology, which is radiation-hard. it contains some 60,000 transistors, forming four high-speed 16 × 16-bit fully parallel multipliers in a 144-pin grid array (PGA) package. One version operates on fixed-point numbers in two's complement form with unsigned or mixed operands, giving two 32-bit outputs in 150 ns. Another version caters for floating-point input and output of 16 bits mantissa and 6 bits exponent, and multiplies in 200 ns in the fall-through mode, although 100 ns is achievable with one level of pipelining. The programmable matched filter of section 2.2 would require 48 of the fixed-point devices in a parallel-pipeline FFT architecture (17). Both versions have a 16-bit test register and extra pins for initial testing and subsequent in-situ confidence checking.

26.4.2.3 *FFT butterfly*
This could be a development of the above 22-bit floating-point complex multiplier, perhaps in $2\,\mu$m CMOS-SOS technology (24). It would be desirable to include sufficient memory on the chip for data reordering. Data recording in an N-point pipeline FFT architecture requires a maximum of BN bits of memory per stage, where B is the number of bits in the real or imaginary data word, and it would be desirable to include this on the chip. A 1024-point transform would require 22,528 bits of storage.

26.4.2.4 *Log-modulus extractor*
Many coherent processes require the logarithm of the modulus of complex data at about 12-bits resolution. A useful development would have a flexible input format up to 16 bits, and provide outputs up to 12 bits of modulus, log modulus and phase.

26.4.2.5 *FFT address sequencer*
The address sequencing for array- and signal-processors can range from integer counting to complex number patterns involved with FFTs. The need to generate both data and coefficient address patterns for transforms up to 64K points in length will require fast parallel architecture efficiently matched to RAM and capable of supporting multiple algorithms for decimation-in-time, decimation-in-frequency and user-derived functions. The only such chip (21) available at present is the AMD 29540 operating at 5 V and configured specifically for use in the AMD byte-slice pipelined processor family (bus-oriented).

26.4.2.6 *Multi-applicable ALU*
Some arithmetic is always required before and after high-speed DSP. An off-chip processor between FFT stages capable of quantising and arithmetic scaling, normalisation of data formats (fixed/FP) etc would enhance throughput. Functions required would be store, add, subtract, and scaling and normalisation of FP number formats; to convert between different FP formats and fixed point and to perform enhanced parallel I/O data path shifting compatible with DSP array techniques. For radar DSP, FP formats of 22 bits (16 + 6) are acceptable (22).

26.4.2.7 *Digital CFAR processor*
Adaptive thresholding, whether accomplished by cell averaging, binary integration or ratio detection uses a common stream of range cell samples held in memory. One cell is compared to those surrounding it by use of specific

algorithms. A single CFAR chip allowing external selection of a particular algorithm would be an attractive proposition (20).

26.5 Architecture and technology integration issues

VHPIC technology requires a different approach to its architecture and integration in systems. It will probably be most used in programmable, flexible processors.

26.5.1 *Architecture*

The Parallel Microprogrammed Processor (PMP) (23) is an attractive architecture for a future radar processor. Its many identical modules, working in parallel and containing input memory, a processing element (PE), working memory and a bus I/O interface, could be implemented in VHPIC. The microcoded controller configures and controls the modules for the desired algorithms and functions according to a main programme memory. The aim would be for a PE to perform 100×10^6 to 200×10^6 instructions/s, with a basic cycle time of 25 ns to 50 ns.

Alternative work has been undertaken by RSRE and ICL (24) to exploit the Distributed Array Processor (DAP) for radar processing. The Mil DAP version has a 32×32 array of PEs: a far greater degree of parallelism than the PMP. The PE is much simpler than that of the PMP, operating as a 1-bit element with a cycle time of about 140 ns. Mil DAP can perform several of the radar processes, providing that signal bandwidths are reasonably low. Although the programming overhead in assembly language is significant, excellent flexibility is afforded. A VHPIC PE could be faster and more complex.

The systolic array (14) is ideally suited to VHPIC, being a regular array of simple PEs and requiring no further memory or a master control unit. As in the DAP, each PE is connected only to its nearest neighbours and each performs the same function simultaneously, but the PEs and interconnections are hardwired. Although the systolic array is therefore not as reconfigurable as the DAP, it can achieve greater throughput in a smaller processor for direct convolution, DFT, matrix-vector and matrix–matrix multiplication, and rank detection. In this country, work has concentrated on arrays of single-bit PEs (25). For example, a direct-convolution chip is being developed, 48 of which could form a programmable matched-filter for TB products up to 64 and bandwidths up to 10 MHz, requiring no other circuitry whatsoever.

26.5.2 *Technology integration issues*

VHPIC chips, whether the all-hardware specialised processors or the more flexible microcoded processors, need attention given to function partitioning and packaging, and hence interconnection. Although pin-to-gate ratio generally decreases with increasing integration, there is even now a need for chip carriers with over 100 pins, and 1000 pins are predicted when board-level systems are integrated. Therefore VHPIC packages must cater for many pins, thermal conduction, stress, ease of replacement, and must minimise both the time constants of the scaled-down interconnect-line geometry and the propagation delays due to interconnection lengths. The new leaded or unleaded ceramic chip carriers (LCCC) and pin grid arrays (PGA) will aid the development of specialised boards for complete system functions. Zero- or low- insertion-force PGA carriers (26) with more than 220 pins require multi-layer pcbs which are capable of maintaining reliable contact across the array. For surface mounted LCCCs, the trend is towards metal-core pcbs which, because of their sandwiched copper-clad invar construction, have an inherent ground plane, good shielding and excellent thermal conductivity (27). Also, because the carriers are surface-mounted, lithography techniques can be used to lay down the multi-layer circuit patterns. Variants of this approach (28, 29) seek to overcome the problems of thermal stress and controlled-impedance I/O lines associated with surface-mounted chips by using ceramic plug-in cards.

Alternatively, a processor, its controller and memory could be modularised in a three-dimensional stack package, thereby reducing both the inter-module I/O connections and wiring length (30). Yet another approach is to stack module substrates, and propagate I/O connections vertically throughout the stack. The whole structure is enclosed in a demountable container and force-cooled. The ultimate packaging technique may be wafer-scale integration, the interconnection of sub-systems on an undiced wafer.

26.6 Conclusions

It can be seen that future, advanced radar processes will be extremely demanding of technology, and many will be too expensive, too large or simply impractical without VHPIC. As well as the functions summarised in this paper, it is important to consider, for example, a complete ADC subsystem in VHPIC for coherent digitisation, instead of a stand-alone ADC.

Not only radar, but other applications such as spread-spectrum communications, require fast, complex circuits. Thus, certain chips will have many applications, and although their development costs are inevitably high, the more that are produced, the cheaper they will be. Clearly there is a bright future for VHPIC.

Copyright © Controller HMSO, London, 1985

26.7 References

1. SCHOENENBERGER, J. G., FORREST, J. R., PELL, C., 1982, "Active Array Receiver Studies for Bistatic/Multistatic Radar", Proc. Int. Conf. "Radar '82", IEEE Conf. Pub. No. 216.

2. WINDRAM, M. D., 1980 'Adaptive Arrays – A Theoretical Introduction', *Proc. IEE*, **127**, Part F, pp. 243–248.
3. WIDROW, B., MANTEY, P. E., GRIFFITHS, L. J., GOODE, B. B., 1967, 'Adaptive Antenna Systems', *Proc. IEEE*, **55**, pp. 2143–2159.
4. REED, I. S., MALLETT, J. D., BRENNAN, L. E., 1974, 'Rapid Convergence Rate in Adaptive Arrays', *IEEE Trans*, **AES-10**, pp. 853–863.
5. HUNG, E. K. L., TURNER, R. M., 1983, 'A Fast Beamforming Algorithm for Large Arrays', *IEEE Trans*, **AES-19**, pp. 598–607.
6. COOK, C. E., BERNFIELD, M., 1967, 'Radar Signals – An Introduction to Theory and Application', Academic Press, New York.
7. TURIN, G. L., 1960, 'An Introduction to Matched Filters', *IRE Trans*, **IT-6**, pp. 311–329.
8. STOCKHAM, T. G. Jr., 1969, 'High-Speed Convolution and Correlation with Application to Digital Filtering', in Gold B, Rader, C. M., 'Digital Processing of Signals', McGraw-Hill, New York Chap. 7.
9. HARTT, J. K., SHEATS, L., 1971, 'Applications of Pipeline FFT Technology in Radar Signal and Data Processing', IEEE Eascon '71 Record, pp. 216–221.
10. BLANKENSHIP, P. E., HOFSTETTER, E. M., 1975, 'Digital Pulse Compression via Fast Convolution', *IEEE Trans*, **ASSP-23**, pp. 189–201.
11. KELLOG, W. C., Nov. 1982, 'Digital Processing Rescues Hardware Phase Errors', *Microwaves and RF*, pp. 63–67, 80.
12. SILVERMAN, H. F., 1977, 'An Introduction to Programming the Winograd Fourier Transform Algorithm (WFTA)', *IEEE Trans*, **ASSP-25**, pp. 152–165.
13. NUSSBAUMER, H. J., 1981, 'Fast Fourier Transform and Convolution Algorithms', Springer-Verlag, Berlin.
14. KUNG, H. T., LEISERSON, C. E., 1980, 'Algorithms for VLSI Processor Arrays', in Mead, C. A. and Conway, L. A. 'Introduction to VLSI Systems', Addison-Wesley, Reading, Mass, Section 8.3.
15. WARD, J. S., BARTON, P., ROBERTS, J. B. G., STANIER, B. J., 1984, 'Figures of Merit for VLSI Implementations of Digital Signal Processing Algorithms', *Proc. IEE*, **131**, Part F, pp. 64–70.
16. THOMPSON, C. D., 1983, 'Fourier Transforms in VLSI', *IEEE Trans*, **C-32**, pp. 1047–1057.
17. JOHNSTON, J. A., 1983, 'Parallel Pipeline Fast Fourier Transformer', *Proc. IEE*, **130**, Part F, pp. 564–572.
18. TSUNEO *et al.*, 1983, 'Circuit Techniques for VLSI Memory', *IEEE J*, **SC-16**, pp. 463–470.
19. GEC Hirst Research Centre, 1983, 'Fixed-Point Complex Multiplier', Advance Information Sheet, MA 754.
20. TRUNK, G. C., 1983, 'Survey of Radar ADT', NRL Report 8698.
21. A.M.D., 1983, 'Array and Digital Signal Processing Data Book', Section 7.
22. VEENDRICK, H. J., PFENNINGS, L. C., 1982, '40 MHz Multi-Applicable Digital Signal Processing Chip', *IEEE J*, **SC-17**.
23. MUEHE, C. E., McHUGH, P. G., DRURY, W. H., LAIRD, B. G., 1977, 'The Parallel Microprogrammed Processor (PMP)', Proc. Int. Conf. 'Radar '77', IEE Conf. Publication No. 155, pp. 97–100.
24. ROBERTS, J. B. G., 1984, 'The Impact of New VLSI Technology on Radar Signal Processing', Proc. Int. Conf. 'Radar '84'.
25. IEE, 1984, 'Systolic Arrays – A Review from First Principles', Colloquium Digest No. 1984/14.
26. WALTERSDORFF, H. R., 1983, 'A Zero Insertion Force Connector for PGAs', IEEE, Proc. 33rd Electronic Components Conf.
27. REYNOLDS, R., STIRLING, K., Feb. 1984, 'Surface Mounting and Leadless Chip Carriers', *Defense Electronics*, **16**.
28. CUMMINGS, J. D. *et al.*, 1982, 'Technology Base for High Performance Packaging', IEEE, Proc. 32nd Electronic Components Conf.
29. DORLER, J. *et al.*, 1982, 'A ceramic card concept', *ibid*.
30. MURASE, T. *et al.*, 1982, 'High Density Three-Dimensional Stack Packaging', *ibid*.

Chapter 27
High performance and high speed integrated devices in future sonar systems

J. M. Beresford
Plessey Marine Ltd, UK

27.1 Introduction

The high performance criteria dictated of future sonar systems is placing increasing demands on technology to realise the more-complex high-performance signal processing functions. Koral (2) stated in 1979 that the performance of large-scale integration (LSI) circuitry had not reached the point where performance was no longer the issue for radar and satellite processors. Now the sonar engineer is also looking to high-performance and high-speed digital integrated devices to achieve systems which would be unachievable without them. For example, applications may be found in adaptive systems, such as might be required for a sonar beamformer, processing an array of hydrophones whose numbers are in the order of thousands. The emerging Very Large Scale Integration (VLSI) technology is partially providing a solution to these processing requirements. Excluded from our consideration is the category of general purpose (Von Neumann architecture) processors which are assumed always to be needed for control purposes but will not match the high-speed requirements of the sonar processor.

Customising integrated circuits for sonar signal processing falls into two categories:

(a) Application-Specific where the architecture and partitioning of the circuit is such that the devices perform dedicated processing for a given task. An example would be a VLSI circuit designed to perform a fast Fourier transform (FFT).
(b) Specific-to-Signal Processing but have applications in sonar signal, processing functions in general

The problems for sonar are that in general the complex systems are few in number and it is difficult to justify financially and resource the design of application specific circuits. According to Bogert and Wetlesen (3), if full use is to be made of VLSI technology, new system architectures need to be developed. These architectures will stand or fall by the matching of the technology capabilities to the architectures. Thus it is not surprising that "systems architects" are now designing the future VLSI chips. This chapter describes the building block methodology whereby VLSI circuits are identified which support an architectural concept designed to provide commonality across a wide range of signal processing functions such as beamforming, filtering and spectrum analysis and common to many sonar applications. In addition the architecture is designed so that future sonar systems may be realised with modules (Function-Units comprised of VLSI circuits) configured using simple guidelines and computer-aided design methodology to minimise the design-resource requirements. This is in contrast to the "modular functional element" architectural concepts proposed by Koral (2) which combined LSI building blocks at a much lower level.

27.2 Function unit architecture

The key feature of the high-speed, high-performance devices is that they should fit into an architecture which enables signal processing systems to be built from signal processing modules termed Function-Units. These Function-Units can be programmed in micro-code to perform a complete signal processing function such as filter, an FFT or correlation. An example of the macro-code instruction set, written at assembly-language level, would be 'floating-point FFT butterfly'. The signal processing functions listed below are sufficient for designing the signal processor for most sonars and once the funtion-units are microprogrammed they may process data on receipt of the appropriate macros defining these functions:

1. Window.
2. Convolve.
3. Correlate.
4. Fourier Transform.
5. Co-ordinate Convert.
6. Divide.

Hence in principle a single Function-Unit could be programmed to accept hydrophone data and perform all the digital sonar signal processing prior to display. However, for larger complex systems it is more practicable to have a number of Function-Units whose hardware facilities are arranged to perform a subset of the total macro set and operate on a limited number of input channels.

27.3 Sonar system architecture

A key feature of the architectural concept is that the Function-Units are designed to be interconnected in a hierarchical way and computer-aided design

234 Integrated devices in future sonar systems

techniques are used to facilitate the sonar-system designer. An example of the hierarchical structure is shown in Figure 27.1.

The approach adopted is that each Function-Unit is interconnected to its neighbours by an input-output (I/O) queue. These interfaces are all identical and each Function-Unit has two-input and two-output ports. The control of the Function-Unit is effected by appending a block header to the input-data stream. The three words are; a tag which identifies the data block, and two control words which identify the routing of the data through the Function-Units and declares the macro(s) to be implemented at each Function-Unit. The appending of the control word is performed by additional circuitry termed 'Control Interface', while the tag would be added at the data source.

The 'Control Interface' contains memory which is addressed and loaded with data by the overall system control (host general purpose micro-computer). Each block of data, input to the 'Control Interface', addresses the memory by means of the tag (channel indentifier) and the control words found at that address are appended to the channel identifier to complete the block header. A 'Control Interface' therefore controls via the block header the routing of data through a number of Function-Units and specifies which macro will be implemented at each Function-Unit. With some constraints relating to input/output port interconnections the structure is hierarchical and a Function-Unit maybe specified with or without the addition of a 'Control Interface'.

27.4 Building blocks to realise function units

The objective was to optimise Building Blocks, which could be realised in VLSI high speed circuitry, and would fit the common architecture designed to interconnect the circuits to form Function-Units. These building blocks required to achieve the Function-Units and perform the macros described in 27.2 above have been rationalised as follows:

1. Input/Output Unit (I/O).
2. Floating-Point Complex-Arithmetic Unit (FPCAU).
3. Flexible Address Generator (FAG).
4. Memory RAM/PROM (Store).

The interconnection of these units in the common architecture enables the number and type of block to be varied to meet a particular specification. To this end the system designer is provided with a computer-aided design (CAD) procedure to assist in the determination of the numbers of circuits to realise given macros for a specified number of data channels. It is highlighted that each Function-Unit could have different combinations of the building blocks dependant upon the sonar specification but the architecture of all Function-Units is identical. Thus our commonality of VLSI chip design and ease of use is established.

Integrated devices in future sonar systems 235

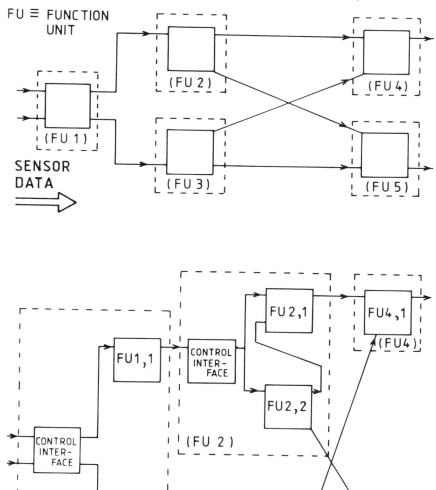

Fig. 27.1. Two levels of description of a typical system showing the hierarchical structure

236 Integrated devices in future sonar systems

In general a Function-Unit would normally contain one I/O Unit to input data, one I/O Unit to output data, Flexible Address Generator (FAG), a memory unit and one arithmetic unit (FPCAU). Nevertheless the size of memory and number of FPCAU's can be tailored to meet the macro and channel capacity. A typical Function-Unit is shown in Figure 27.2 showing the five VLSI Building Blocks and two medium scale integration (MSI) blocks.

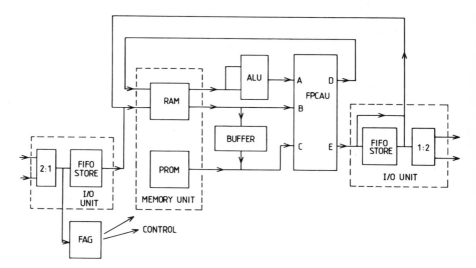

Fig. 27.2. A typical function-unit

The I/O Building Block is a queue comprising a FIFO store which transfers data using a handshake protocol between output and input ports of two interconnecting Function-Units, or from the input port of a Function-Unit to the input port of the same unit via the memory block.

The FPCAU (Figure 27.3) has three complex data input ports, two complex data output ports and can be programmed under instruction of a control word to perform the following signal processing operations in floating point arithmetic:

1. Complex multiplication – for bandshifting and windowing.
2. Complex multiplication and accumulation – for FIR filtering and beamforming.
3. FFT butterfly.
4. Co-ordinate conversion.
5. Divide.
6. Second order IIR filter point.

Appendix A provides a more detailed description of the arithmetic unit.
The FAG consists of two parts, a flexible address generator and a sequencer.

Integrated devices in future sonar systems 237

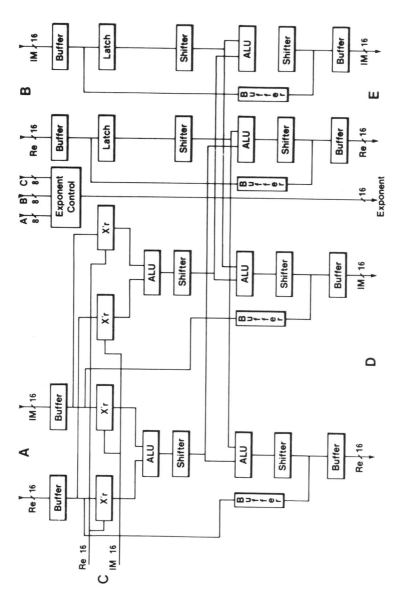

Fig. 27.3 A floating point complex arithmetic signal processing unit

The former part contains counters for controlling sequencies, number of steps and so on for the various signal processing functions, while the sequencer contains the microcode written in PROM to realise a given set of macros. Each time a new block of data is input to a Function-Unit via the I/O Unit FIFO queue, the block header is decoded to cause an interrupt to the sequencer which proceeds to initiate a sequence of macros. For each macro, the sequencer controls the flow of data between the various hardware blocks until finally after completion of the macro, the processed data is located "blocked up" in the output I/O Unit.

The Fourth Building Block Memory, is necessary for the storage of data and coefficients to realise the macros. The size of this memory and its organisation is tailored to meet the channel capacity and type of macro specified of the sonar system. This Building Block is considered to be obtainable from commercially-available VLSI devices.

27.5 VLSI implementation

The implementation of Function-Units using commercially-available TTL devices from AMD and TRW have been realised to prove the architectural concepts. The throughput rate of the FPCAU building block implemented in TTL may be specified by quoting that a non-pipelined radix-2 FFT butterfly may be realised in 200 nsec, albeit at tens of watts power dissipation and occupying a densely packed extended double Eurocard. The aim is to use the emerging CMOS or Bipolar technology to achieve the specification shown in Table 27.1.

Table 27.1 *Estimated performance of VLSI building blocks*

Building block	Speed	Gate count
Floating point complex Arithmetic unit (FPCAU) 16 bit mantissa, 8 bit exponent	100 nsec for an FFT Butterfly	50,000
Flexible address generator (FAG)	100 nsec per New address	30,000
Input/output unit	100 nesec per I/O operation	30,000

The Table shows the three aforementioned Building Blocks each realised in a single VLSI chip with a pin out exceeding one hundred, but at a power dissipation of less than one watt (cf the commercially available equivalent).

It is anticipated that these gate counts and speed of operations will evolve as silicon technology advances. The design of these chips will be accomplished by utilisation of current research technology and by the use of the extensive design tools being developed jointly by industry and the information technology programmes. These include the design of circuits using cell libraries, hardware simulators, and silicon compilers.

The CAD methodology however, does not comprise only tools at the VLSI chip design level. It will be recalled that the architectural concept called for easy use of the VLSI circuits when configured to Function-Unit level.

Computer-aided design: Simulators

To build Application-Specific VLSI chips without simulators can lead to costly mistakes. For the proposed general-purpose VSLI Building Blocks, once working devices have been established, simulation is also required to ensure that the desired accuracy in the arithmetic processing is achieved and to check that the overall system transfer function is obtained with adequate throughput rates. The simulators in addition will allow quick interaction with the system designer to study the performance of different sonar specifications or different signal processing techniques/implementations, for example, changes to a filter coefficient.

The simulation of a system, comprised of Function-Units, to check accuracy and system transfer function is relatively simple due to the realisation with one type of arithmetic unit operating at known rates and number of operation cycles to realise each of the well defined macros. Thus the simulator is able to quantify the accuracy of the processing and throughput rates given the organisation of the Function-Units, the macros used and the input sampling rates.

The three levels of simulator design are:-

1. VLSI chips:- described in Section 27.5 above.
2. Function-Units:- the simulation of the interconnection of the VLSI chips to verify speed/performance.
3. System level:- the simulation of the interconneciton of the Function-Units to establish a sonar design from a high level specification.

The design procedure masks the sonar designer from the detail of the VLSI chip sets. He will use the system level simulator (3. above) interactively to obtain a first-off configuration of the Function-Units, he would then move to the lower level simulator (2. above) to check on detailed accuracy/performance. This simulator could for speed advantage use a hardware Function-Unit as a pheripheral to the general-purpose computer simulator.

In summary the impact of the high-performance high-speed device once established into a working Function-Unit Signal Processing Architecture and supported with simulator software, is to provide the sonar designer with a similar capability to that obtained with a general-purpose microprocessor architecture in terms of ease of use and transparency of the circuity. The one exception is that the Function-Unit processor will operate at speeds 2-3 orders of magnitude faster than the general-purpose microprocessor architecture.

27.6 Acknowledgements

This chapter is published by permission of Plessey Marine Ltd and extracts from original material published in IEE proceedings. Wadham (1).

The author would like to acknowledge D. Wadham and his co-workers K. Lamb, J. Read and L. Williams for their major contribution in the concepts and worked described.

27.7 References

1. WADHAM, D., April 1984, 'The Modular Design of Sonar Systems Using Functional Units built from a standard set of VLSI Building Blocks'. *IEE Processings* Vol. 131, Part F. June 1984, Issue 3.
2. KORAL, W., 1979. 'The Impact of Advanced LSI on Future Signal Processing Architecture' *IEE Conference Publication Number 180* Case studies in Advanced Signal Processing, pages 244–247.
3. BOGERT, H. and WETLESEN, M. December, 1982. *Dataquest Markets Section Volume 1*, Chapter 2 Section 2.11.

27.8 Appendix A – Floating point complex arithmetic unit (Chip)

With reference to Figure 27.3, each of the words A, B, C, D and E are 40-bits wide made up as a floating-point representation as follows:-

Two × 16-bit mantissas for real and imaginary parts, and an 8-bit exponent which is common to both and scales each equally. The binary coding is two's complement in all component parts.

The general arrangement for the ports is A and B reserved for input, D and E reserved for output and port C is for coefficient (except for cross correlation). In addition a control is required. The arrangement as shown would require a device with over 200 pins if no multiplexing of ports were realised. Thus a practical VLSI device may invoke trade-offs between multiplexed ports and packaging technology.

Signal processing

The input complex words, A, B and C in general produces two complex words D and E at the output. An example of a typical arithmetic operation would be a radix-2 FFT butterfly (decimating in time)

$$E_n = B_n + C_n \times An$$
$$D_n = B_n - C_n \times An$$

and another would be complex multiply accumulate

$$E_n = \sum_{n=1}^{N} C_n \times An$$

Combinations of these basic operations enable the realisation of a wide range of arithmetic operations. These include Heterodyning, Beamforming, Filtering (both IIR and FIR), DFTs, FFTs, Co-ordinate Conversion, Correlation, Convolution, Averaging, Normalisation and Division (using Newton-Raphson Iterations).

Chapter 28

Ada

M. J. Pickett
CAP Industry Ltd.

28.1 Introduction

Ada is a new programming language developed under contract for the US Department of Defense. It was procured and designed as, initially an alternative, and eventually a replacement, for the excessively large number of languages and dialects in use in the US defence industry. Its intended field of application is "embedded computer systems" by which is understood systems where the computer is a component rather than a tool in its own right. Such systems include radars, communications equipment, missile-guidance systems, fire control and navigation aids. From 1st January 1984, all new USDoD projects of this kind are required to use Ada as the programming language.

Signal processing is an essential ingredient of many embedded computer applications, and therefore it must be presumed that signal processing software will be expected to be written in Ada. It must be borne in mind that Ada is not a special-purpose language. It is a general-purpose language with characteristics intended to enhance its value for use in embedded-computer systems. Therefore, this paper will highlight some of those characteristics so that potential users may be better able to assess the applicability of Ada.

This paper does not assume the reader to have any knowledge of Ada, nor does it attempt to teach the language. There are already many books on Ada; a few key books are listed in the bibliography.

28.2 Aspects of the language

There is a growing tendency to criticise Ada for the inadequacy (more often lack) of its implementations. In short, because it is new, there are very few compilers for the language, and those that exist will benefit from further development. Notwithstanding this criticism, it is the intention here to concentrate on aspects of the language itself, only referring to implementation in the abstract where that has a bearing. A potential user of the language must also consider the relative qualities of the different implementations of the language before making a choice.

The following four aspects of Ada are examined below: scope, ergonomics, reliability and speed.

28.2.1 *The scope of the language*

By "scope" is meant "what can be done with the language". This can best be discussed in terms of individual features of the language, although Ada has one feature which effectively makes for an open-ended scope: packages.

Ada is a procedural language like the bulk of current programming languages and is therefore directed towards processes which may be described sequentially. But, in addition, it provides facilities for describing parallel processes known as tasks. In Ada, a task is usually a substantial piece of program which may be executed in parallel with other tasks. Communication between tasks is provided for by a rendezvous mechanism whereby one task calls a named entry point of the other (as it might call a procedure) which accepts the call. The calling task is held in the rendezvous until released by the called task. During the rendezvous, information may be exchanged through parameters. In general, the first task to reach an intended rendezvous is held up until the other arrives and the rendezvous is completed, but timeouts and conditional calls and accepts are provided for the avoidance of deadlocks.

A compiled Ada program automatically includes a scheduler for its tasks. Additionally, tasks can be attached to hardware interrupts. Thus an Ada program can be written such that there is no need for a supporting operating system. It is therefore a useful language for programs which must run in bare machines.

The arithmetic features of the language are fairly basic in respect of the functions provided, but much thought has been put into the specification of how values are held and the accuracy of computation. In particular, by writing algorithms which rely only on the minimum properties guaranteed, a programmer can be sure of portability between conforming implementations.

Three kinds of numbers are supported by the language: integers, fixed-point numbers and floating-point numbers. The language itself does not prescribe what ranges of these numbers must be supported, and it is left to the implementor of the compiler whether or not double or multi-length working is provided.

Although the language has no more built-in functions than the average pocket calculator, the package facility is expected to be used to provide libraries of functions applicable to any and all applications.

One other aspect of arithmetic is that results are automatically checked to be within the range defined. That is, for each variable of a program, there is a fixed range of values determined by the programmer. Any attempt to assign to a variable a value outside that range will be trapped automatically by the program.

The input/output facilities of a language are often neglected or are hopelesly inadequate. Ada's came in for so much criticism that they were completely redesigned. The facilities defined for Ada are provided through the package mechanism and it is obviously open to an implementor to provide additional facilities by the same means. Ada input/output is of three kinds. The first kind is for textual information which is serial and line oriented. Simple formatting operations are provided including conversions between internal and external representations. The second kind is for non-textual information and would normally operate with files or through channels accepting and delivering blocks of data. Operation can be serial or direct access where appropriate, direct access being by record number. The third kind of input/output is a low-level facility which is designed to be used to send and receive control signals and to put and fetch data on busses. This might typically be used in tasks servicing interrupts.

Although Ada is a high-level language and does not provide the programmer with the ability to manipulate bits, it recognises that, in embedded systems, the compiler cannot always be allowed to decide how it will use the hardware. Therefore, Ada allows the programmer some control over the packing of data and the location of data and program in available storage. This is a particularly important facility in respect of an ability to handle interrupts.

The package facility is the primary means for extending the usefulness of Ada. In simple terms, a package is a named collection of data and subprograms which may be incorporated into a program. However, the characteristics of the language ensure that packages are integral parts of programs, with consistent and protected interfaces. Packages can be written in such a way that they are reusable by other programs, defining the characteristics of the data objects they manipulate (e.g. complex numbers) or by being parameterised so that different programs may use them to manipulate different kinds of objects.

When Ada was conceived, it was never intended that it would replace Fortran and COBOL in the scientific and commercial communities. Nevertheless, through the use of well designed packages, Ada has been found to be quite usable in both these fields, although it is highly unlikely that the very large bodies of users of both these languages will ever be converted to Ada.

28.2.2 *The ergonomics of Ada*

Among the requirements for Ada was a statement that the new language should be readable and that this was more important than that is should be easy to write. The justification for this statement was that a piece of program is read many times but it is only written once.

Whether or not a program in Ada is easier to read and follow than it would be in some other language is questionable. Certainly, in absolute terms, a programmer can produce fairly opaque programs in Ada, but there are a number of factors which can assist him to express a program quite coherently if he so chooses. Such factors are also an aid to writing the program.

Ada does not constrain the programmer to use short, cryptic names for his objects. The use of meaningful names, which can be as long as a line, is encouraged. Such names can be structured to indicate where in the program they are declared, but to avoid excessive writing, local aliases can be introduced.

Any construct in Ada, other than the simplest statements and declarations, has a defined terminator. Generally this is **end** followed by another keyword specifying the construct being terminated (eg **end case**, **end loop**, **end if** and **end record**). Other constructs may have names assigned to them by the programmer, the more obvious examples being subprograms (procedures and functions), packages and tasks. Blocks and loops also may be named. In these cases, it is usual to terminate the construct by **end** followed by the name assigned to the construct. Although Ada programs may be written to a fairly free format, it is usual to write them such that nested constructs are progressively indented. This technique, together with the construct terminators can make the structure of a program comparatively easy to visualise.

The programmer is able to extend the scope of the built-in functions to operate on data of his own design. For example, he can write functions which perform arithmetic on complex numbers and name these functions with the usual arithmetic operators. He can then write expressions of complex variables in a natural way. Of course, there is nothing to prevent him from extending the operators to have the most unexpected effects, but that would be perverse.

28.2.3 *Reliability*

It is a fundamental requirement of embedded systems that they be reliable. There is no programmer around to sort them out if they go wrong. Unreliability can arise from several sources: the program may contain errors; the system may have been incorrectly specified; some associated hardware may malfunction leading to incorrect data or erroneous computations.

In general, when a fault occurs, it may not be particularly important to identify the cause but it is important that the system survive and continue to operate. This requires faults to be detected as soon as possible and their effects to be contained.

Understandably, a lot of attention has been paid to these requirements in the design of Ada. Ada employs a technique know as strong typing. The programmer is encouraged to consider the kinds of data he is using and how they are to be manipulated, and then to specify their characteristics by means of distinct **type** declarations. For example, he may have velocities which are always in the range 0 to 100 kph which he holds to three significant figures, and also distances and elapsed times with corresponding ranges and precisions. These are regarded as distinct from the built-in real numbers although derived from them. He may arrange that he can divide distances by elapsed times to get velocities, multiply velocities by times to get distances, and divide distances by velocities to get times. The compiler will then allow him to write expressions involving such functions, but it will disallow any attempt to add a distance to a velocity (although they are both basically real numbers). The compiler also builds in checks as part of the program to ensure that the ranges specified by the programmer are not exceeded. Thus the compiler traps some errors during compilation and others during testing of the code.

The compiled checks are normally retained as part of an operational program. When a run-time check fails, an exception is said to be raised and control is passed to a handler written as part of the program. The programmer may also include checks of his own and have them handled by the same mechanism. An exception handler may be written at any level within a program and is the primary means of detecting and containing an error. If a handler is unable to fix up a problem at any particular level, it may pass the exception up to a handler at a higher level. It is only if no handler is prepared for the exception that the program may be said to have crashed.

The safety mechanisms of the language are primarily aimed at detection of errors which appear through apparently erroneous data. Although such effects may often be due to logic errors, the Ada package facility is expected to be used for the development of tried and tested reusable components. By this means, the proportion of original code in any program can be considerably reduced and with it, the scope for logic errors.

28.2.4 *Speed of execution*

It is frequently argued that high-level languages are bound to produce programs which execute more slowly than the equivalent programs hand coded in machine code, and that effect is usually found in practice. Although a major factor is the quality of implementation of the compiler, it is also a fact that often the compiler has to allow for special cases which the low-level programmer knows can never happen. Ada attempts to relieve the compiler of the need to consider special cases by requiring the programmer to define his data more precisely (strong typing), by restrictions (eg a function may not return values through its parameters although a procedure can), and by specifying certain things to be erroneous (eg unprotected changes to shared variables by parallel tasks).

Although speed is important in embedded systems, Ada appears to favour safety to speed and hence the level of checking it employs. If he believes he can do without them, the programmer can suppress the compiled checks, whether on selected variables or on all variables of specified types. But even with the checking present, an efficient compiler can often eliminate unnecessary checks. For example, if the programmer indexes an array with a variable of the type he defined for the array's index, the variable is bound to be within range and no check is necessary at this point.

If speed is important, it cannot be too strongly emphasised that benchmarking should be used as one of the means for selecting between available compilers.

28.3 Ada availability

Ada is a new language. It is large and compiler development for it is a major investment. The number of serious compilers under development is probably of the order of ten, but for most of these it has been a struggle. Therefore it is to be expected that early versions of these compilers will not be top quality products. In particular, at the beginning of 1984, only one serious compiler had succeeded in passing the suite of validation tests instituted by the USDoD. Nevertheless, many organizations are already using the incomplete compilers which are still under development. One producer claims to have over 450 compilers in the field.

By the end of 1984, perhaps two or three more compilers will have been validated, but most of the announced compilers are not expected to seek validation until late 1985.

28.4 Summary

This paper has identified some of the aspects of Ada which should be of interest to a programmer choosing a language for a signal processing application. No attempt has been made to suggest that Ada is the ideal choice, but it is recommended that it be considered.

28.5 References

1. BARNES, J. G. P., 1984, 'Programming in Ada', Addison-Wesley, London, England.
2. DOWNES, V., and GOLDSACK, S. J., 1982, 'Programming Embedded Systems with Ada', Prentice-Hall International Inc, London, England.
3. ICHBIAH, J. D. et al., 1983, 'Reference Manual for the Ada Programming Language', United States Department of Defense, Washington, D.C., U.S.A.
4. NISSEN, J., and WALLIS, P., 1984, 'Portability and Style in Ada', Cambridge University Press, Cambridge, England.
5. NISSEN, J. C. D., WICHMANN, B. A., et al., 1982, 'Ada-Europe Guidelines for Ada Compiler Specification and Selection', NPL Report DITC 10/82, National Physical Laboratory, Teddington, England.
6. PYLE, I. C., 1981, 'The Ada Programming Language', Prentice-Hall International Inc, London, England.

Ada is a registered trademark of the U.S. Government, Ada Joint Program Office.

Chapter 29

Future for cryogenic devices for signal processing applications

S. A. Reible
Lincoln Laboratory, Massachusetts Institute of Technology,
Lexington, Massachusetts 02173–0073, USA

29.1 Introduction

The principal advantages of superconductive circuits in both analog and digital applications are high switching speed (hence large bandwidth), low propagation loss and low power dissipation. In spite of rapid advances in semiconductor and other competing technologies, superconductive circuits continue to maintain a nominal order-of-magnitude advantage in speed. Superconductive circuits operate at very low power levels, but the need to provide a cryogenic environment eliminates most of their advantages when comparing total power requirements. However, the low power dissipation translates to a potential advantage in circuit density.

The distinct speed advantage of superconductive technology will find its greatest usage in real-time signal processing. Potential applications include: correlation and analysis of wideband signals, matched filtering, and integrated-focal-plane-array processing. Bandwidths of 10 GHz and signal processing gains of 1000 and beyond are projected for superconductive correlators and other signal-processing devices. Direct signal processing of gigahertz-bandwidth signals will allow very high resolution and possibly even three-dimensional details to be determined for identification and classification purposes at high frame rates. Furthermore, because many future systems will require cryogenic sensors (e.g., for mm-wave or infrared radiation), the superconductive circuits can be integrated into the sensor module with little additional investment in power or space. Hybrid analog/digital signal-processing architectures implemented with cryogenic circuits will extend the available processing gain in noisy environments.

This paper will describe some of the essential functions required for analog and digital signal processing. Cryogenic circuits can fulfil all of the required functions for signal processing, often with more than an order of magnitude increase in bandwidth capability over that which is available from existing room-temperature circuits.

29.2 Analog signal processing

To perform analog signal processing it is necessary to configure a structure or a collection of components whose equations of motion duplicate the desired signal-processing function. Such integrated components must simultaneously perform multiple functions such as delay, distributed multiplication and summation with often very stressing requirements (1). A list of these functions, along with the most stressing requirements and the superconducting component which can fulfill the particular function, is given in Table 29.1. Some of the functions, notably multiplication and readout, may be provided by cooled semiconductor circuits with the potential for higher dynamic ranges but at the cost of greater cooling requirements.

Table 29.1 *Functions required for analog signal processing*

Function	Requirement	Component
Delay	Low dispersion Loss low Compactness	Stripline
Tapping	Accurate weights	Proximity coupler
Multiplication	Adequate dynamic range	Mixer
Spatial summation	Phase coherence	Microstrip
Time integration	Adequate storage time	Resonator
Readout	Sensitivity	Logic

The most significant figure of merit for the signal processing system is processing gain. This is defined as the improvement in the signal-to-noise ratio of the output signal relative to the input signal. The maximum potential processing gain of a device is equal to its time-bandwidth (TB) product, and in well-engineered integrations of device and peripheral electronics this maximum processing gain can be realized to within about 1 dB.

One of the most stressing requirements of analog signal processing is the need to provide for temporary storage of several thousand wavelengths of analog signal without excessive phase or amplitude distortion. Acoustic technology has filled a substantial niche in signal processing applications because acoustic waves can provide low-loss delays with small phase distortions (2). Superconductive technology can provide low-loss transmission lines on dielectric substrates with much shorter delays than available from acoustic devices but with at least an order of magnitude increase in bandwidth (1). The loss per wavelength of delay for both structures is plotted in Fig. 29.1 as a function of frequency. Projected losses are shown for Rayleigh waves on lithium niobate at

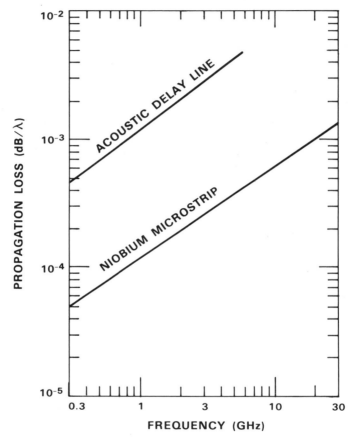

Fig. 29.1 *Propagation loss per wavelength of delay for acoustic and superconductive delay lines*

300 K and for niobium microstrip lines at 4.2 K on 25-μm-thick, low-loss substrates. Comparable losses with normal metal transmission lines at 4.2 K would be about three orders of magnitude larger. The same superconductive technology provides dense interconnects for signal transmission in digital circuits.

A pulse expander/pulse compressor has recently been realized at Lincoln

Future of cryogenic devices for signal processing applications 251

Laboratory (3) by integrating superconductive transmission lines and proximity couplers on a 5-cm-diameter silicon wafer, as shown in Fig. 29.2. The tapped-delay-line structure currently provides a signal processing bandwidth of 2.3 GHz and a TB product of 86 with measured amplitude ripple of 1.5 dB and phase deviations from quadratic of 9 degrees rms. The compressed pulse for a Hamming-weighted linear-frequency-modulated chirp waveform is shown in the inset in Fig. 29.2; it has a respectable peak-to-side-lobe level of 27 dB. Side-lobe levels of -40 dB are expected in refined devices.

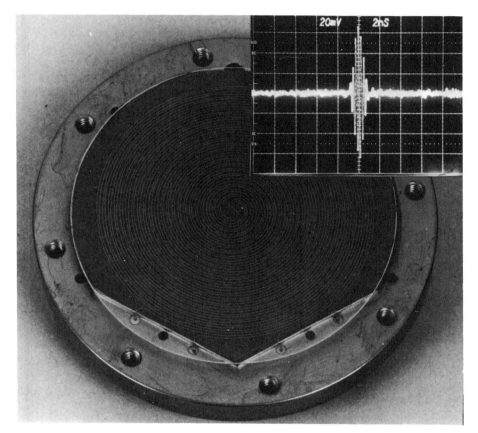

Fig. 29.2 *Photograph of superconductive tapped delay line fabricated on 5-cm-diameter silicon substrate with Hamming weighting. Inset shows compressed pulse output of this 2.3-GHz bandwidth, 37.5-ns dispersion device*

By adding to a tapped delay line structure both mixers for multiplication and a microstrip line for spatial summation, the generic functions of convolution or correlation can be realized. Such a programmable device is being developed at Lincoln Laboratory (4). A photograph of a superconducting convolver is shown

252   Future of cryogenic devices for signal processing applications

in Fig. 29.3. A meander line provides delay, 50 proximity couplers provide sampling, 25 junction-ring mixers provide local multiplication and two short transmission lines provide summation (or integration) and output. This device has recently been demonstrated with both gated-cw input tones and wideband ($\sim$ 1 GHz) chirped waveforms. The wideband output response is shown in the inset in Fig. 29.3. The high spurious levels are due to the formation of undesired products in the mixing interaction. With a dynamic reference, these devices can provide programmable matched filtering. Alternatively, a time integrating correlator can be realized by replacing the summation output lines with individual superconducting resonators, and by using superconductive logic to provide the readout function. Signal processing bandwidths of 10 GHz and TB products of 1000 and more are projected for both the passive and programmable superconductive devices.

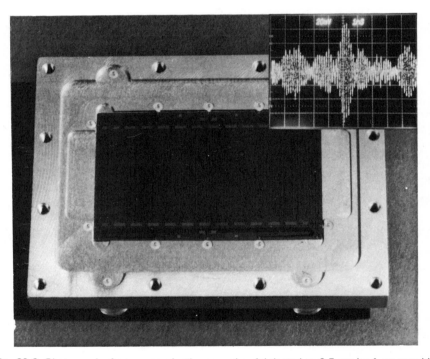

Fig. 29.3 *Photograph of a superconductive convolver fabricated on 2.5-cm by 4-cm sapphire substrate. Inset shows wideband ($\approx$ 1.6 GHz) compressed pulse output*

Although the time-integrating devices provide a convenient time buffering of the data, many analog signal processing devices transform signals from one parameter range (e.g., frequency) to another (e.g., time) without providing any significant reduction in data. The output of devices such as the pulse compressor and convolver must be sampled at their full bandwidth and decisions be made

Future of cryogenic devices for signal processing applications 253

with high-speed digital logic. Circuits to perform these functions can be provided very conveniently by superconductive technology. Several research laboratories (5), (6) are developing analog/digital (A/D) converters with superconductive circuits. Three-bit accuracies with about a 0.5-GHz analog bandwidth have been demonstrated, with the primary limitation being the lack of an adequate sample and hold.

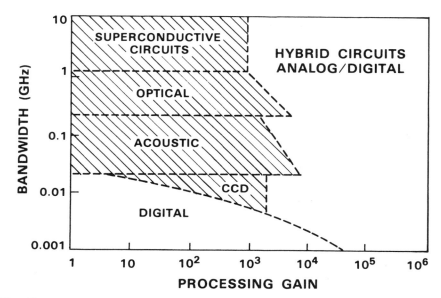

Fig. 29.4 *Projected bandwidth versus processing gain regimes of analog, digital and hybrid signal-processing systems*

The projected bandwidths for several analog signal processing devices are shown in Fig. 29.4. Charge-coupled devices (CCDs) occupy a bandwidth regime which is being compressed on the bottom by digital systems and on the top by acoustic devices. Acoustic devices for signal processing exist in well-engineered forms and currently operate near their projected bandwidth limits of several hundred megahertz. Optical and superconductive devices exist in a less advanced state of development, but the superconductive devices already provide considerably larger bandwidths than available from acoustic devices. The signal processing gain available from most of these analog devices will range from somewhat under 10^3 to 10^4.

29.3 Digital signal processing

In the past year we have seen the discontinuation of several large commercial efforts to build digital circuits and a general purpose, mainframe computer with

Josephson technology. A development effort at IBM extended over more than a decade and brought forth many exciting new developments in technology, circuit concepts, cryogenics and systems engineering (7). The Josephson technology itself failed, largely through a lack of useful power gain, to provide adequate design margins for the essential high-speed memory with the required extremely low error rates. This should not be viewed as a fundamental limit, but rather an inability to reach the commercial marketplace in a timely fashion. Other cryogenic concepts and technologies may one day provide this gain and a superconductive mainframe computer effort may be reinstated.

Turning our attention to the requirements for signal processing, we note that these applications do not, in general, require very low error rates but rather demand extreme speed for highly structured data flow. Stringent environmental requirements must also be met. Superconductive technology today can clearly supply the speed and reliability with niobium-based designs. Environmental considerations, such as space and power requirements, depend more on the development of an adequate cryocooler technology than on the device technology itself.

The most frequently encountered function in radar and communications is matched filtering. It is instructive to consider the digital fast Fourier transform algorithm (FFT) as a means to achieve programmable matched filtering. It is possible to quantify the magnitude of the process in terms of the number of arithmetic operations per second (ops). Digital signal processing techniques are highly developed and considerable savings can be realized by employing FFT techniques. But the required computation rate for programmable matched filtering (8) is still at least $20 \, B \log_2 TB$. Dedicated digital processors can currently provide computation rates of about 2×10^9 ops. Based on this rate, the current bandwidth limits for digital signal processing systems were sized as a function of processing gain and plotted as shown in Fig. 29.4.

Note that in order for signal processing bandwidths to increase by an order of magnitude, computation rates must increase at least as fast. This places a very real limit on the bandwidth capability of digital signal processors. Conversely, processing gain, dynamic range and accuracy can often be more easily extended in a digital system than in an analog signal processor.

What is the potential for increasing the digital computation rate in the coming decade? In spite of very high speed logic ($< 40 \, \text{ps}$), the projected computation rate for the Josephson general purpose computer was less than 10^9 ops. This is in large part because the computation rate tends to be limited by propagation delays, clock skews and reset times rather than by the logic switching speeds in both room-temperature and cryogenic systems. Parallel processing systems allow the designer to substantially increase the computation rate. This technique is commonly implemented in dedicated semiconductor processors but at a considerable cost in terms of power and space requirements. Because of the extremely low power requirements ($\approx 2 \, \mu\text{W/gate}$) of superconductive logic, specialized superconductive systems such as array processors could be achieved

Future of cryogenic devices for signal processing applications 255

with extremely dense gate configurations and considerably higher computation rates.

29.4 Hybrid signal processing

Many applications, such as signal analysis and matched filtering, require very large signal processing bandwidths. Because of the bandwidth limitations of digital processors, these needs, currently and well into the future, will only be met by analog signal processing devices.

However, in a hybrid (analog/digital) system, an analog device could provide the capability of large processing bandwidth, while digital circuits would extend the available processing gain. Such an implementation has recently been demonstrated with acoustic convolvers (9). Here the convolver preprocessed a 100-MHz waveform, while a binary integrator extended the processing gain.

It is interesting to note that superconductive technology can provide all the elemental analog functions required for signal processing, and all the digital functions as well. This one technology, unlike any other, is therefore fully capable of implementing entire hybrid signal processing architectures. The extended signal processing capability which may be provided by future hybrid systems (superconductive and multiple technology configurations) is indicated in Fig. 29.4.

29.5 Discussion

The digital superconductive research activities which were recently curtailed in North America were directed toward the development of high-speed logic and memory circuits for high-performance mainframe computer applications. While the potential market volume is large, it may not have been the most suitable choice for existing cryogenic technology. It is common to judge a new technology in terms of its potential to replace a well-developed technology in existing applications. This seldom happens, even if considerable effort and financial resources are expended on developing the new technology. Rather, the new technology usually finds its market niche in rapidly developing new applications.

Exciting new opportunities exist for cryogenic devices in signal processing applications. Superconductive devices for pulse compression and convolution have recently been demonstrated and are being further developed. These analog devices offer bandwidth capabilities which are unreachable with other technologies. Yet in applications such as signal sorting and image processing, a real need exists for extremely large bandwidths. Digital circuits fabricated with the same technology can provide very high speed circuits for sampling and data sorting. Hybrid systems implemented with cryogenic circuits may offer the user unparalleled capabilities in signal processing in the future.

29.6 Acknowledgements

The author gratefully acknowledges R. S. Withers and A. C. Anderson for use of their results on the tapped-delay-line structures and for their collaboration in developing fabrication techniques, and R. W. Ralston and J. H. Cafarella for their advice and guidance.

29.7 References

1. REIBLE, S. A., 1982, 'Wideband Analog Signal Processing with Superconductive Circuits', *1982 Ultrasonics Symp. Proc.*, 63–74.
2. WILLIAMSON, R. C., 1977, 'Measurement of the Propagation Characteristics of Surface and Bulk Waves in $LiNbO_3$', *1977 Ultrasonics Symp. Proc.*, 323–327.
3. WITHERS, R. S., ANDERSON, A. C., WRIGHT, P. V., and REIBLE, S. A., 1983, 'Superconductive Tapped Delay Lines for Microwave Analog Signal Processing', *IEEE Trans. Magn.*, **MAG-19**, 480–484.
4. REIBLE, S. A., 1983 'Superconductive Convolver', *IEEE Trans. Magn.*, **MAG-19**, 475–480.
5. DHONG, S. H., JEWETT, R. E. and VAN DUZER, T., 1983, 'Josephson Analog-to-Digital Converter Using Self-Gating-AND Circuits as Comparators', *IEEE Trans. Magn.*, **MAG-19**, 1282–1285.
6. HAMILTON, C. A. and LLOYD, F. L., 1981, 'Design Limitations for Superconducting A/D Converters', *IEEE Trans. Magn.*, **MAG-17**, 3419–3419.
7. ANACKER, W., 1980, 'Josephson Computer Technology: An IBM Research Project', *IBM J. Res. Develop.*, **24**, 107–112.
8. CAFARELLA, J. H., 1983, 'Wideband Signal Processing for Communications and Radar', *Proc. Nat. Telesystems Conf.*, 55–58.
9. BAKER, R. P. and CAFARELLA, J. H., 1980, 'Hybrid Convolver/Binary Signal Processor Achieves High Processing Gain', *1980 Ultrasonics Symp. Proc.*, 5–9.

Chapter 30
Application of the ICL distributed array processor (DAP) to radar signal processing

J. F. Cross, BSc, M. M. Hayward, MA,
I. D. Longstaff, PhD, C Eng, MIEE
Royal Signals and Radar Establishment, Malvern, UK

30.1 Introduction

RSRE are sponsoring the development of a fast and powerful signal processor based on the architecture of the ICL Distributed Array Processor (DAP). A number of application areas exist where a suitable signal processor is critical to continued success. Possible examples may be found in the fields of image processing, antenna array processing and multi-mode pulse Doppler radars. This latter application is being used as a 'pace setter' to establish the design and performance criteria which must be met. It is believed that the variety of tasks, the data rates, the computational load and programmability required for this application will demand an array processor solution which will be suitable for the many other applications.

This paper describes the evolution of the ICL DAP from its original implementation in MSI technology, when it was incorporated into the architecture of a large mainframe computer, to the ruggedised 'MIL' DAP currently under development in LSI technology. This will function as a compact stand-alone signal processor suitable for the multi-mode radar. It should perhaps be mentioned however that the DAP is not the only solution and in the UK individual radar manufacturers are pursuing their own architectures, which also show great potential.

30.2 Historical background

The first customer ICL DAP (the MSI DAP) was set up in Queen Mary College in the University of London in 1979 and a national service is offered on it. Since then similar systems have been installed at other Universities and research

258 Application of the ICL (DAP) to radar signal processing

establishments in the UK. The MSI DAP consists of a large array (64 × 64) of processors and is hosted by a large ICL 2900 mainframe. There are well over 200 registered users on the QMC service and the machine is fully utilised. Initial applications tended to fall into the category of large regular number crunching work but now successful work is being carried out in a wide variety of other areas which were not considered at the time of the early hardware design.

The main language used is a special version of FORTRAN (called DAP FORTRAN) with extensions which exploit the array architecture of the DAP. The design of this language has influenced the development of the international standards for the FORTRAN array extensions (ANSI FORTRAN 8X). Fine tuning of DAP performance can be achieved by assembler subroutines. In addition, the MSI DAP user has at his disposal all the ICL 2900 mainframe software tools such as editors, file systems with interactive program development. The MSI DAP allows multiprogramming, thus fast experimental jobs can be quickly turned round while, at the same time, a quota of longer jobs (some very long) are kept steadily humming away in the background.

The national DAP service at QMC does not, however, just supply the hardware and software wherewithall to allow use of the DAP. A special applications oriented support unit was set up in 1979 (funded jointly by industry and Government) to help the research community to make use of a machine with such a novel architecture. A measure of the wide breadth of the active research can be gauged by the following list of topics which arose at a recent symposium (reported in the ICL Technical Journal, May 1983):

1. Mote Carlo methods.
2. Explicit finite difference calculations (plasma equilibrium for toroidal configuration).
3. Molecular dynamics calculations.
4. Non-linear econometric models.
5. Numerical optimisation.
6. Networking problems.
7. Variable precision arithmetic (search for prime numbers).
8. Processing of map data (Ordinance Survey).
9. Compiling of high level languages.
10. Image processing.
11. Sea models.
12. Parsing of English text.

In all the above examples the DAP was being used not merely for its own sake but because it provided a powerful research tool. No generalisation can be drawn (either from the above list or from any list of current research items) about types of problems which are particularly suited, except that in many cases DAP performance was found to increase steadily with the scale and complexity of the problem.

A number of advances accrued from the decision to host the MSI DAP with

Application of the ICL (DAP) to radar signal processing

an ICL 2900 mainframe, such as guaranteed access to a sophisticated communications network, use of a large reliable filestore and a fast processor for interactive programme development. However, two drawbacks emerged: a few applications (image processing particularly) quickly found they became IO bound; a remote user on a terminal would not expect to achieve real time display or update of his stored DAP image. The second drawback relates to the size and cost of the MSI DAP and host system, which was outside the budget of small research groups. Both these problems are addressed by the smaller, cheaper LSI DAP which is the subject of the programme under consideration.

To round off this section on the historical background a brief description of the architecture of the MSI DAP will be given. As mentioned earlier, it is essentially an array of 64 × 64 processing elements under the control of a central master control unit (MCU). In determining a match between the target applications and a regular cheap form of hardware, an early decision was made that each individual processing element (PE) should be extremely simple. Thus the PEs are bit organised with each PE having available to it a few thousand bits of fast random access store. The totality of store of all PEs is called the array store. The PEs allow simple arithmetic and logical operations to be carried out with the data taken from the PE or its neighbours. Instructions are broadcast from the MCU so that simple matrix operations are achieved by PEs operating simultaneously and in an identical fashion. A program dependent mask on the array can be used with little additional overhead to inhibit operation of selected PEs. Other options include the fast placement of data (saved, say, in MCU registers) into the array store in regular form and, conversely, the condensing of logical matrices in the array store into MCU registers. Although the PEs operate, for simplicity, at a single bit level, a ripple carry is supplied between PEs as part of the neighbourhood operations. Other features to be found in conventional architectures are also, of course, supplied (such as procedure calling, jumping, conditional tests and indexing).

The cycle time of the MSI DAP is 200 ns. A better feel for the performance which this cycle time signifies may be given by quoting the time for a 64 × 64 32 bit floating point matrix multiply (30 ms) or a full pivoting 64 × 64 matrix inversion (50 ms). These assume DAP FORTRAN programming and better times would be achieved with assembler code. However, the real proof of the performance benefits of the DAP come from users accounts of solutions to real problems. At the symposium mentioned above, many users reported results of comparative runs on DAP and on other machines acknowledged to be in the highest class of computing power. There was certainly a feeling that the DAP was in the same league and that for certain problems could be rated as the world's best performer.

In order to understand why RSRE decided that ICL should be approached with a request to develop their MSI DAP into a radar signal processor, it is necessary to look very carefully at the requirements and in particular at the very special design considerations needed when such signal processing is executed on an airborne or agile platform.

260   Application of the ICL (DAP) to radar processing

30.3 Requirements for signal processing for agile platform airborne radars

Because of the limitations imposed by fairly small agile airframes, much of the radar system is compromised by available space, available prime power, allowable weight, competition for real estate with other 'equally' important sensors and so on. At a very early stage in the evolution of a new weapon system many of the parameters of the radar system are fixed within quite fine limits. It is up to the radar designer to provide as much radar capability within the allocated package. Arguably the most promising and exciting avenue opening up to provide more and more cpaability is to look toward future semiconductor advancement and be in a position to take advantage of the high throughput, small size, low power consumption technologies which are promised to be 'around the corner'.

If the 'rest' of the radar has sufficient flexibility to allow transmission and reception of a menu of waveforms then the single radar system can be reconfigured to support many different roles such as air-to-air search, air-to-air combat, navigation aid through radar imaging and air-to-ground weapon delivery. As far as this paper is concerned, the main differences in these modes are:

(a) The radar waveform.
(b) The received signal environment.
(c) The processes required to provide the required information.
(d) The presentation of the information.

It has been tempting in the past to identify the requirements of a few of the possible modes which will exercise store requirements, arithmetic throughput (complex multiplies for example), data dependent operations and logical operations. Using these ideas, it is possible to arrive at required figures for instructions or operations or complex operations per second. From experience, however, those concepts can lead to a great deal of confusion because it is very difficult to compare one man's ops with another, no matter how much care is taken. Confusion is probably least when comparing architectures which have special purpose pipelined arithmetic. In this case, it is sensible to identify how many arithmetic operations per second are required to perform an FFT for example. This can then be easily mapped on to the clock cycle requirements of the arithmetic pipeline or perhaps the degree of parallelism required. However, in considering architectures where the arithmetic is built up in software at the bit level, the situation becomes very unclear. For example, it has been demonstrated that a machine with a superficial real multiply rate of 20 MHz can approach an effective multiply rate of a factor of five higher when performing FFTs.

FFTs are only one example of algorithm running on a radar job and are often followed by a data reduction process (a detection process) and a correlation process (decoding ambiguities).

With these difficulties in mind, RSRE has chosen to examine the various facets of the architecture by benchmarking a set of algorithms which are representative of a particular mode and which will exercise the architecture's abilities in the following areas:

(a) Method of input and output of data and how this process reflects on the concurrent processing.
(b) The ability of the architecture to handle highly structured data sets and provide arithmetically intensive processing on such sets, eg multiple FFTs.
(c) The ability of the architecture to communicate with 'adjacent' radar cells and using collaborative assessments to set thresholds in difficult 3-D environments to form detection decisions.
(d) To take all candidate detections and confirm and decode to produce target information such as range and Doppler. This latter type of process is not at all arithmetic intensive but rather relies on rapid data reorganisations to accomplish correlation or on an ability to form lists and rapidly assess these lists.

The extent of reconfigurability is an interesting compromise. With today's technology there seems little doubt that if some special purpose hardware can be included it will be beneficial in some processes, yet ideal in others. Our firm belief is that the machine should be flexible by virtue of starting with completely flexible hardware which is configured by software.

The high dependence which such a programme would have on the software element in turn imposes a requirement on the research and the implementation development environment. It became important to have an algorithm development environment which allows the radar engineer on the one hand to work in a research mode (where new algorithms and control mechanisms are created and evaluated) and on the other hand in production mode (where large complex software systems are constructed, updated and maintained). This implies the existence of specialised software tools many of which are common to any large software based project but some of which need to be geared to the special characteristics of radar signal processing and of parallel processing.

30.4 The military DAP solution

The discussion given in the previous section represents the views of the small group at RSRE which is concerned with advancing the state of the art for lightweight multi-mode airborne radars. These are generally pulse Doppler radars with the ability to undertake a number of functions within the general area of detection, tracking, and ground mapping. Each mode requires different signal processing algorithms operating with high data and computational rates.

The DAP architecture is attractive as it offers a sufficiently flexible 'general purpose' machine whose architecture matches the overall problem; that is identical processing in each resoltion cell of the radar with an ability to perform efficient local computations and to produce (where applicable) overall resumés.

262 Application of the ICL (DAP) to radar signal processing

A feasibility study was placed on ICL to look at the architectural changes required to the MSI DAP architecture and to bench mark the representative set of algorithms, the objective of which were described in Section 30.3 above. This feasibility study was followed by an outline design which resulted in the specification of a product which is called the Military DAP and is now under development.

The Military DAP has an architecture very similar to the MSI DAP described above except that the array is one quarter the size (32×32) and the necessity for a mainframe host is removed. New technology has allowed a raw performance improvement and, of course, a size reduction. Also stand-alone operation is possible; alternatively the Military DAP can be 'hosted' by the ICL PERQ, a small scientific workstation. The IO limitations on the MSI DAP are overcome by a new programmable fast input device which uses a corner turning and reorganising buffer and achieves a rate of 40 Mbytes a second, with one cycle stolen from the DAP for each plane of data transferred. The DAP system (including the code store and the optional host interface) occupies 16 PCBs (each PCB being 4 times the size of a Eurocard) and with the cooling and power supply it occupies a volume of $53 \times 51 \times 30$ cm. The machine consumes 750 W, weighs 64 kg and has an operating temperature range of -20 C to 30 C. The suite of program development tools which are available parallels that for the MSI DAP, ie DAP FORTRAN and assembler. The PERQ system operates in a UNIX environment for algorithm development. Full execution time tracing and diagnostics are avaialble as on the MSI DAP but the high resolution screen on the PERQ workstation is likely to make this even more attractive and user friendly than the remote terminal servie on the MSI DAP.

The cycle time of the military DAP is 145 ns, which allows the full medium PRF search algorithm to be completed well within the dwell time of the scanning radar beam.

This algorithm operates on a schedule of 5 PRFs with an average of 100 range cells per PRF. Data from each range cell is in the form of a 10 bit in-phase and a 10 bit quadrature component. Processing takes the time sequence of 64 complex samples from each range cell, applies a weighting function and then a 64 point discrete Fourier transform (DFT). The log modulus of each spectral component is then computed. This process takes a total of 8 ms for the 500 range gates, ie an average of $16 \mu s$ for each weight-DFT-log-modulus computation.

The next stage in the processing is to test the signal amplitude in each cell of the range-Doppler maps (500×64 cells) against a threshold derived from the local area average. This constant false alarm receiver (CFAR) detection process takes 4 ms.

Finally an N/M detection test is applied across the 5 PRFs after resolving all possible range-Doppler ambiguities. This process takes up to 1 ms, depending on the number of targets, making the total duration of the complete algorithm 13 ms.

Application of the ICL (DAP) to radar signal processing   263

Each section of the algorithm exploits different (but overlapping) aspects of the DAP architecture and it is partly for this reason that the algorithm is seen as a good 'pace setter'. The windowing and DFT exploit the parallelism and the ease with which constant scalar multiplications can be optimised in a one bit architecture. The CFAR exploits the DAP array connectivity (and, naturally, the parallelism) whereas the anti-aliasing and target detection exploits a host of Boolean, lookup and counting operations well suited to the logical nature of the DAP processing elements.

The military DAP is currently under production, with the first batch of working prototypes and algorithms available in December 1985. As well as being used in the multi-mode pulse Doppler environment as a 'pace setter', these machines will also be used in the other fields mentioned in the introduction, that is image processing, classification and antenna array processing.

30.5 Future development

The very regular nature of the DAP architecture makes it an ideal candidate for VLSI design technology. It becomes advantageous to exploit the cheapness of the silicon at high levels of integration and at the same time to ask for design changes which, although not altering the overall architectural concepts, allow improved performance over and above that achieved by a mere decrease in feature size. Design changes in the area of the array of PEs and store make the greatest impact on the total performance. There are at least seven such design changes currently under consideration. These range from the use of a multi-bit processing element to the incorporation of the main memory on the chip containing the processing elements. There is a choice between the various modes of packaging of store and PE as well as between the different silicon technologies. These choices, when considered alongside the above design change choices, result in a large number of potentially useful VLSI options. Detailed analysis of these choices is a very important on-going task and feedback from the militatry DAP programme is having an important influence on that analysis by supplying a very specific but typical reference point. A few of the sample conjectured VLSI machines which have been the subject of the detailed analysis include the following: an extremely small DAP which would have the power of one half of the Miltiary DAP; a DAP with a very large array store size which is very much smaller than the Military DAP but 4 times the performance and, finally, a DAP a third to a quarter the size of the Military DAP but thirty or forty times more powerful.

With continued success in the Military DAP programme it is expected that the evolution of this generic form of architecture will go hand in hand with VLSI advances to provide one or more of the above options, each of which would have its own fairly broad class of application. Such applications in the defence area would include advanced seekers for the extremely small DAP; 'intelligent'' image processing and surveillance systems for the very large array store version; and programmable signal processors for multimode radar and sonar systems using the very high performance version.

Copyright © Controller HMSO, London 1985.

Chapter 31
Advanced image understanding
Andrew C. Sleigh, MA

31.1 Introduction

Image understanding is concerned with the important area of interpreting the output of a sensor and initiating an appropriate response to the perceived scenario. This might mean rejecting a component on a production line because of some defect, selecting the highest value target accessible to a missile, or sending a meaninful message down a narrow bandwidth channel. Current device and software technology will to enable very high-performance computers to have widespread use in a vast range of new products. In most cases this computing power will need to interact directly with the environment rather than through a keyboard input, and image understanding techniques provide the mechanism to bypass this human bottleneck. These techniques must be capable of representing a large body of knowledge and experience in order to cope with the variability and complexity of un-constrained environments. This Chapter discusses the main difficulties in achieving this aim, and describes a system which is being built to demonstrate the feasibility of some of these techniques.

Note that the use of the word "image" is convenient because of its relationship to our visual sense. However, any data vector can be treated as a multi-dimensional image and what follows is equally applicable to radar signals, sonar data or any other input derived from the environment.

31.1.1 *Main problem in image understanding: combinatorial explosion*
A typical sensor, for example a TV camera, is capable of producing several megabytes of data every second. Even a single black and white frame is capable of portraying $10^{1,000,000}$ different patterns. Even if a we suppose an object within a binary image covers a 50 × 50 patch, then the number of distinct patterns still exceeds 10^{75}. Most of these patterns are either equivalent or are physically implausible, and the essence of image understanding is recognising which patterns are different in a way significant to the system's response, which are equivalent, and which need not be considered for solution. There are some cases where this can be achieved by an isomorphic transform, where the nature of the

266   Advanced image understanding

problem is such that similar situations always cluster in a given region of the isomorph. Examples of this occur in analysing satellite data where the different spectral reflectivities can be used to identify crops, urban areas, etc by rotation of the multi-spectral data and projection onto some decision axis. But in most problems such a convenient transform does not exist, and full use of knowledge of the world must be used to contain the combinatorial explosion which is implicit in the dimensionality highlighted above.

The traditional approach has been to extract a small number of meaningful features (eg size, colour, object aspect ratio) from the scene and compare these to the values obtained when training the system with a set of sample situations. To succeed it is important to choose the features carefully so that they are reliable and statistically independent with respect to the training data, and to have a representative set of training samples. The weaker the features are in discriminating the classes of situations, more features are needed and a larger training set is needed for a given level of discrimination. The size of the training set soon becomes a fundamental limit on the complexity of the situation which can be handled, with the result that training is feasible when there are 10 or so features and about the same number of different classes. Being able to analyse such a small number of features requires that these be simple properties (such as the size of an object), and this necessarily means first separating the object from its background, a process known as segmentation. Since this must be done before the object has been identified it is notoriously unreliable unless the object is in a very clear background, even when local constraints can be imposd by statistical methods such as relaxation.

To avoid the limitations imposed by this unstructured statistical approach, methods are needed to represent and analyse concepts which relate more closely to the abstract significance of a situation. The next section discusses two aspects of this problem: representation of knowledge derived from human experience, and the use of problem solving strategies which can reduce the size of the search involved in inferring conclusions from this knowledge.

31.2 Declaritive methods in unconstrained image understanding

31.2.1 *Declaritive representation instead of statistical clustering*

To avoid the impractically large data sets needed to train statistical-decision processes, it is possible to use human expertise to define the properties and relationships of objects or situations. For example, instead of gathering many photographs of different specimens, one could define a car as "an object which has wheels in contact with the ground and between two and five doors" (plus some other rules). Not only is this very much easier to do, but we know the system will recognise *any* car with these properties, not only those it had been shown in training. By setting up a database of knowledge about the occurrences and relationships within the image understanding system's field of discourse, we can construct an arbitrarily rich set of distinctions without the explosive growth in a training database. In some instances no training data whatsoever may be needed.

Representing this expertise, which may be extensive, must be by a convenient and transparent process. Formalisms which allow the concise expression of such relationships have existed for a long time in the branches of mathematics related to formal logic, but recently the advances in computing techniques has made the application of this type of mathematics to real problems a feasible proposition. One method is to implement rules such as those in the car example in the form of a production system (Nilsson (1)). This is a general form of knowledge representation, which encompasses formal syntactic analysis, where the satisfaction of a particular condition in the current state of affairs of the system causes an action which modifies the state of affairs in some way. In formal syntactic analysis these "productions" correspond to the re-write rules which define the grammar, in less formal systems the rules can assert or retract facts into the knowledge base, or they can cause some action in the outside world, for example moving a robot arm. Computer languages which allow effective writing and implementation of production systems are quite well established, and computer architectures which can efficiently run these languages are not far away. A trivial example of a production system to identify edges could be:

IF left is black, right is white THEN context = edge, orientation = 1
IF context is edge, region (orientation) is uniform, region (orientation + 4) is uniform THEN context = confirmed edge.

A practical system would have many rules and constructs to help control the application of the rules.

Another closely related method of knowledge representation available is that based upon predicate logic and logical resolution, perhaps the best known being the PROLOG computing language (Clocksin and Mellish (2)). PROLOG is an implementation of first-order predicate calculus with linear input resolution of Horn clauses, and represents facts as predicates such as 'likes(janet, john).' and relationships such as 'friends(X, Y):- likes (X, Y), likes (Y, X).', which says that two people are friends if they both like each other. "Goals" can be set which the PROLOG system will attempt to satisfy by a depth-first search through the latent decision tree formed by the relationships, and the satisfaction of a goal can be used to cause some external action. The use of a language such as PROLOG allows complex relationships to be concisely expressed, and transparently modified even during run time. An example of how this can be applied to recognising shapes consider:

rectangle(Pta,Ptb,Ptc,Ptd,Sureness):-
rightangle(Pta,Ptb,Ptc,Str1),rightangle(Ptb,Ptc,Ptd,Str2),
rightangle(Ptc,Ptd,Pte,Str3),nearto(Pta,Pte,Separation),
boxstrength(Str1,Str2,Str3,Separation,Sureness)

where the predicates rightangle, nearto and boxstrength are assumed. PROLOG will attempt to satisfy the above rule for every set of points which exist in the database.

31.2.2 Inference control to avoid exhaustive searches

The use of production systems and predicate logic provide a means for representing knowledge and drawing inference, but they do not in themselves shed any light on how to do this quickly with finite computing resources. Any real-time system which has a large amount of input data to process and a large number of potential relationships to analyse, needs very powerful hardware. More importantly it needs to follow some problem solving strategy designed to draw the right conclusion from the data (at least most of the time) without explicitly considering every possible outcome. Clearly humans reason in this way: we approach a problem on the basis of our knowledge of what methods worked before, we attempt to pick out rules which exclude possibilities, and we use analogies to guide our search for a solution. These problem-solving strategies are taught to us during our education (some may be in our DNA), and we have a varying but generally limited ability to devise new strategies during our lifetime. Any attempt at un-constrained image understanding must be able to carry out this process if it is going to have any practical use, and this area must be considered a key area of research for all autonomous systems.

Whereas we already have quite powerful tools to represent knowledge (such as PROLOG), methods for representing and enforcing problem-solving strategies are extremely weak. At the moment there are three types of strategy which can be used: structuring and partitioning the problem domain; the use of contexts, models and hypotheses; and varying the basic inference mechanism. These are described below. It should be noted that we have no powerful and elegant ways of implementing these strategies, and ad-hoc and generally unsanitary methods must be used, eg the "cut" and "repeat" operations in PROLOG. A breakthrough in this area is desperately needed.

Problem structuring is used to divide knowledge and relationships into different processes on the assumption that some sub-goal exists which can be proven using a known subset of the total knowledge. For example, it is possible to extract edges and lines from an image without needing to know that tables have four legs or that pigs do not fly, and isolating these relationships from each other greatly reduces the combinatorial search implied. For example, the VISIONS system (Hanson and Riseman (3)) breaks the analysis into hierarchical processes progressively finding Edges, Vertices, Regions, Surfaces, Volumes, Objects, and Schemes, with each stage interpreting the output of a previous level using knowledge local to that operation.

When it is not possible to divide the problem domain a-priori, models and contexts can be used to interactively vary the order in which knowledge is applied. For example, if the problem is to look for a man, instead of analysing all line segments in an image against the knowledge about all objects, one could consider only line segments whose position and length conform to a model of a man. At a lower level one might use a model of a right angle to select particular pairs of lines. Models are useful when, as is often the case, the number of possible interpretations is smaller than the amount of input data. When this is not the case models are still useful if "cues" can be found which can sufficiently narrow the range of options.

Contexts can be used to select a sub-set of the system knowledge on the basis of conclusions already inferred or by the use of cues which may be generated by relatively crude methods. As an example, one would expect to search the knowledge base differently if the scene is countryside rather than urban; or if an image contains many long straight lines one might look first for polygonal shapes rather than conic sections.

Finally the inference operation itself may be altered. For many problems, the searching of a decision tree is seldom best carried out on a depth-first basis (as does PROLOG). Breadth-first or minimum-cost to solution are examples of strategies which generally find a solution more rapidly. The inference process can proceed from goals towards the input facts, or it may work from the data towards goals, or some combination of both evaluation techniques.

The problem solving strategies can be used to reduce drastically the amount of computation required, but generally with the risk of excluding some solutions. Hence the methods must be used with care to only exclude improbable situations, or else be used in a way which might only delay a solution if an incorrect intermediate assertion is made. This highlights one difference between logic and artificial intelligence: logic is concerned with always finding solutions to a problem if any exist, without regard to how much effort is involved; artificial intelligence is concerned with making useful decisions within a limited period, even if these decisions are sometimes not the best possible.

The next section shows how these methods have been applied at RSRE to demonstrate their use in image understanding problems.

31.3 Example system: the RSRE "ASARS" image understanding system

The ASARS system is a methodology for general-purpose image understanding. Its distinguishability features are the use of models at all levels in its processing hierarchy, and its use of application-independent knowledge bases at the low and intermediate processing levels. The simplified structure is shown in Figure 1 (Sleigh and Hearn (4)).

The first stage in the processing (written in POP) uses models of line segments to assert into a database the presence of all conforming segments. The detail of the model cannot be described here, but essentially contains a set of common-sense conditions which must be satisfied if a particular point in an image is to be connected to a neighbour. One of these conditions relates to the variation in

270   Advanced image understanding

local brightness around each point, and this is assessed by invoking lower level models of edges, bars, fuzzy boundaries and texture which use a simple production system of rules about local variations in contrast to ascribe the appropriate local property. Examples of this database of line segments is shown in Figure 2, while Figure 3 shows how these lines are formally represented in the PROLOG data base which is passed to the next process.

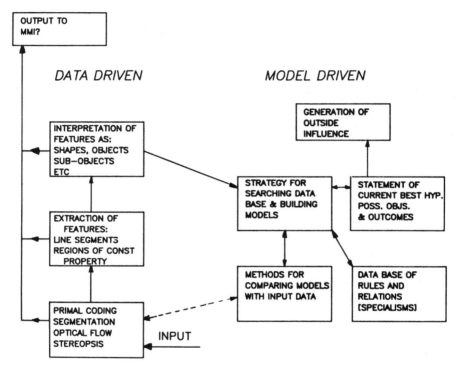

Fig. 31.1 *Processing philosophy of the ASARS image understanding system*

The second stage takes the database of line segments and asserts basic shapes which are identified using models expressed as a set of relations in a PROLOG inference system. If these models are expressed simply, as in the example at the end of section 31.2.2, all the rectangles in the database of Figure 3 will be correctly found, but the inefficient depth first search of PROLOG takes over 45 minutes on a VAX 11/780. It has therefore been necessary to add a control strategy which groups the line segments on the basis of their connectedness and analyses each group as a separate context. This reduces the evaluation time to less than 1 minute, and the shapes asserted by this process are shown in Figure 4. Higher-level models are then applied to form further assertions. The partially obscured rectangle in Figure 4 can be recovered by a relation which links shapes

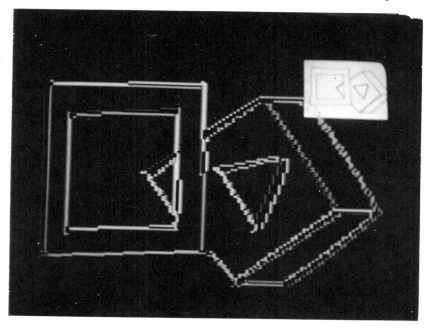

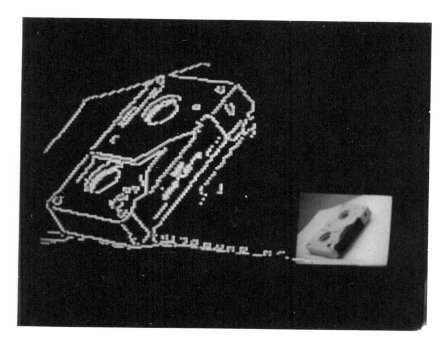

Fig. 31.2 *Examples of line segment extraction from images*

272 Advanced image understanding

with three sides at right angles to an appropriately sited right angle or pair of lines; the arrangement of the rectangle and two parallelograms sharing common lines can be used to assert the presence of a rectangular box; and in more general situations combinations of shapes can be grouped to form elements such as hand shapes, wheel shapes, window shapes, etc. The important attribute of these

```
/*   rect10protpl  */
/*   parameters are:   */

/*  (line no., real ends, fitted ends, strength, angle, length)  */

line_seq(1,[65,59],[66,93],  [65,59],[65,93],16,180,34).
line_seq(2,[64,92],[63,60],  [64,92],[64,60],18,0,32).
line_seq(3,[79,30],[65,45],  [79,30],[65,45],21,223,21).
line_seq(4,[77,107],[87,106], [77,106],[87,106],16,90,10).
line_seq(5,[57,76],[57,83],  [57,76],[57,83],18,180,7).
line_seq(6,[56,51],[56,36],  [56,51],[56,36],18,0,15).
line_seq(7,[78,84],[67,58],  [79,84],[67,58],22,335,29).
line_seq(8,[79,78],[85,53],  [79,78],[85,53],21,13,26).
line_seq(9,[50,63],[55,75],  [50,63],[56,75],23,153,13).
line_seq(10,[102,75],[77,106], [102,75],[76,105],21,221,40).
line_seq(11,[73,109],[66,95], [74,109],[66,95],23,330,16).
line_seq(12,[87,108],[75,109], [87,108],[75,109],17,265,12).
line_seq(13,[56,75],[56,55],  [56,75],[56,55],16,0,20).
line_seq(14,[101,73],[111,73], [101,73],[111,73],18,90,10).
line_seq(15,[110,75],[104,75], [110,75],[104,75],17,270,6).
line_seq(16,[113,76],[89,108], [113,76],[89,108],18,217,40).
line_seq(17,[91,27],[114,75], [91,27],[115,75],18,153,54).
line_seq(18,[80,27],[90,27],  [80,27],[90,27],17,90,10).
line_seq(19,[65,41],[79,27],  [65,41],[79,27],21,45,20).
line_seq(20,[64,22],[64,41],  [64,22],[64,41],16,180,19).
line_seq(21,[21,22],[63,21],  [21,23],[63,21],13,89,42).
line_seq(22,[18,97],[20,23],  [19,97],[19,23],16,0,74).
line_seq(23,[65,94],[19,98],  [65,94],[19,98],18,265,46).
line_seq(24,[63,22],[21,25],  [63,24],[21,25],14,269,42).
line_seq(25,[21,26],[20,95],  [21,26],[20,95],13,181,69).
line_seq(26,[21,96],[63,93],  [21,96],[63,93],15,86,42).
line_seq(27,[67,93],[73,104], [67,93],[73,104],21,151,13).
line_seq(28,[74,105],[100,73], [74,105],[100,73],19,39,41).
line_seq(29,[99,72],[80,30],  [99,72],[79,30],23,335,47).
line_seq(30,[82,30],[100,72], [81,30],[101,72],21,155,47).
line_seq(31,[90,30],[83,29],  [90,29],[83,29],16,270,7).
line_seq(32,[111,71],[91,31], [111,71],[91,31],19,333,45).
line_seq(33,[111,76],[112,72], [111,76],[112,72],13,14,4).
line_seq(34,[88,105],[110,77], [88,105],[110,77],18,38,36).
line_seq(35,[57,52],[58,76], [57,52],[58,76],10,178,24).
line_seq(36,[26,35],[56,33], [26,36],[56,33],14,84,30).
line_seq(37,[26,87],[26,36], [26,87],[26,36],13,0,51).
line_seq(38,[57,84],[27,87], [57,84],[27,86],16,266,30).
line_seq(39,[55,36],[28,38], [55,36],[28,38],15,266,27).
line_seq(40,[27,39],[27,83], [27,39],[27,83],15,180,44).
line_seq(41,[28,84],[56,82], [28,84],[56,82],14,86,28).
line_seq(42,[56,81],[47,60], [57,81],[47,60],24,335,23).
line_seq(43,[48,59],[55,52], [48,59],[55,52],20,45,10).
line_seq(44,[70,59],[78,78], [69,59],[78,78],21,155,21).
line_seq(45,[84,53],[71,59], [84,54],[71,59],19,249,14).
line_seq(46,[67,57],[84,52], [67,57],[84,52],19,74,18).
line_seq(47,[86,53],[79,83], [87,53],[80,83],21,193,31).
line_seq(48,[56,54],[50,62], [56,54],[50,62],13,217,10).
line_seq(49,[65,46],[65,58], [64,46],[65,58],13,175,12).
line_seq(50,[63,59],[63,46], [63,59],[63,46],18,0,13).
line_seq(51,[57,33],[57,50], [57,33],[57,50],15,180,17).
line_seq(52,[63,45],[63,23], [63,45],[63,23],19,0,22).
```

Fig. 31.3 *Database of facts automatically produced from the line drawing in Fig. 31.2*

primitive shapes is that they should not need a scene specific context in their definition. This has processing speed implications, exploits commonality across applications, and enables special processors to be developed economically. However, this restriction need not preclude selective context dependent analysis being performed to confirm the existence of a particular hypothesis.

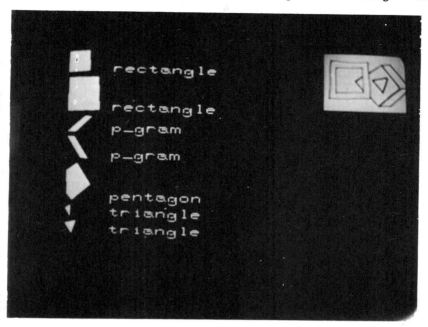
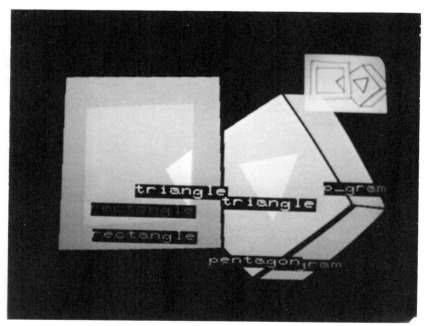

Fig. 31.4 *Reconstruction of the shapes found from the line facts in Fig. 31.3 by the "Shape Expert"*

274 Advanced image understanding

The third stage is a complex process which takes the object primitives asserted by the second process and attempts to form an interpretation which is consistent with all other information to hand. This is the most complex area where structured and "blackboard" systems, fuzzy logic and automatic concept formation will all play a part.

In addition to developing a methodology and advanced techniques, an important part of the ASARS programme is a complementary development of processing hardware to perform these operations in real time. To this end a multi-processor system known as DIPOD is being built by Logica for RSRE which will have a processing power equivalent to about 10 VAX 11/780's in the first instance, extendible by at least another order of magnitude. This has many hardware features needed to run languages such as POP and PROLOG efficiently and will be able to generate a line segment database as least as complex as that in Figure 3 at TV frame rates.

31.4 Conclusion

Representation of knowledge derived from human expertise coupled with suitable problem solving strategies provides a method for achieving advanced image understanding in complex and unconstrained scenes. The example system being built at RSRE has demonstrated the automatic conversion of a TV image into a database suitable for a shape finding expert system, and has shown how such an expert system can be controlled to achieve analysis speeds which are close to real time. Although the interpretation examples have so far been simple, the methods have clear extension to more complex shapes and relationships. With the coming development of special inference engines, the feasibility of using a knowledge based approach for the real-time interpretation of sensor outputs looks very promising.

31.5 References

1. NILSSON, N. J., 1982, Principles of Artificial Intelligence, Springer-Verlag.
2. CLOCKSIN, W. S., and MELLISH, C. S., 1981, Programming in Prolog, Springer.
3. HANSON, A. R., and RISEMAN, E. M., 1978, 'Visions: A Computer System for Interpreting Scenes', Computer Vision Systems, Academic Press, 303.
4. SLEIGH, A. C., and HEARN, D. B., 1984, 'Description of a Method for Finding Generalized Shapes in Images', RSRE Memorandum 3683.

Copyright © Controller HMSO, London 1985

Chapter 32
Advanced signal processing aspects of automatic speech recognition

J. S. Bridle

Joint Speech Research Unit, Princess Elizabeth Way, Cheltenham, UK

32.1 Introduction

Someday machines that recognise speech will be commonplace. People will talk to computers, typewriters, toys, TV sets, household appliances, cars, door locks and wristwatches (White [1]). Before we look at the signal processing and pattern processing that will be the basis of the next generation of automatic speech-recognition systems, we shall consider the nature of the signal that these devices will have to deal with.

32.2 Stages in speech communication – from mind to mind

We are so familar with using speech that is often difficult to remember that the whole process is immensely complex and poorly understood, despite the efforts of scientists from many disciplines, such as acoustics, neurophysiology, linguistics, computer science and even signal processing.

Normal human speech is a means of communication between minds. As the message passes between two minds it takes many forms. Most of the stages involved are in the brains of the speaker and the listener and very little is really known about the form the message takes there. However, we can investigate the way that speech sound-pressure waveforms are launched by the vocal tract, and transduced by the ear.

32.2.1 *Speech production*
The complex, constantly-changing pattern of sound that carries most of the information is produced by the interaction of a wide-bandwidth source of sound (produced in the larynx, at a constriction, or by the sudden release of air pressure) and the frequency-selective action of the vocal tract, which depends on

its shape. The vocal tract is shaped by the articulators, which include the tongue and the lips.

32.2.2 Speech waveforms

The telephone system is based on the observation that speech communication is possible (with various limitations) if we measure the sound-pressure waveform near the speaker's mouth and reproduce an approximation to it near the listener's ear. For telephone-quality speech, we need to reproduce the first few kilohertz of the audio spectrum, with a signal-to-noise amplitude ratio of at least 100:1. This needs tens of thousands of bits per second (say 50 Kb/s), but can be reduced to about 10 Kb/s with compromises in quality, by using special techniques (Holmes [2]).

32.2.3 Speech perception

The patterns of sound pressure entering the ears are transformed by an exquisite natural signal processing system involving hydromechanical resonance and many stages of neural analysis. In the early stages the main action is to lay out the pattern as a function of time and frequency.

32.3 Levels of description – from waveform to meaning

Speech can be described at many levels, ranging from the shape of waveforms to the meaning of whole sentences. Speech science is largely concerned with identifying useful levels of description (Repp [3]) and relating descriptions at different levels. Speech technology typically deals with methods for moving between levels automatically.

As children we learn that words are made up from a limited set of "sounds", which occur in a different sequence in each acoustically-distinct word. The sounds correspond roughly to letters of the alphabet (very roughly in English). These basic sounds are the phonetician's phonemes. What could be more natural to analyse, synthesise, recognise and transmit speech signals in terms of phonemes? Unfortunately, life is not that simple.

Speech scientists now recognise that there is no simple one-to-one correspondence between the linguistic "sounds" (phonemes) and physically-measurable sounds (which are discrete neither in time nor in acoustic properties). The patterns of speech sounds are not like beads on a string: it is rather as if the phonemes were eggs placed on a conveyer belt, then crushed between rollers!

A basic problem, then, is the gulf between the analogue world of SIGNALS (continuous, flowing patterns of sound whose properties depend on the speaker, the context and many other factors) and SYMBOLS, which computer programs might relate to meaning or conventional text. Kopec et al. [4] have pointed out that the intrinsic difficulty is made worse by a communciation gap between those experienced in signal processing and those experienced in symbol processing.

There has been some success in synthesising intelligible speech from symbolic specifications, including conventional text, but all practical, working automatic speech-recognition systems avoid bridging the signal-to-symbol gulf, and leap from sound patterns to the identification of whole words or phrases.

32.4 Speech signal processing – from waveform to pattern

"Raw" speech waveforms are not suitable for speech-recognition purposes. The reason is not so much that the data rate is very high, but that there is an enormous amount of variation in waveform detail from utterance to utterance of the same word, even for the same speaker. "Vocoders" are speech communication devices which attempt to reproduce not the waveform itself but important properties of the pattern. Vocoders need only a few thousand bits per second, and exploit the so-called "carrier nature" of speech (Dudley [5]). The carrier is the broad-band excitation of the vocal-tract, and the main linguistic message "modulates" this with a time-varying pattern of resonances.

One vocoder analyser principle that is also in use in automatic speech recognition is the filter-bank analyser. A bank of 10 to 30 band-pass filters are each followed by rectifiers and low-pass filters, which remove the short-time structure due to the vocal tract excitation. The resulting pattern can be sampled at a rate corresponding roughly to the rate of movement of the vocal tract, say 40 to 100 "frames" per second.

The two main alternative vocoder principles, which have also been used for speech recognition, attempt to model the pattern of vocal tract resonances more directly. In "linear-prediction analysis", (Markel and Gray [6]) the parameters of a mathematical resonance model are fitted to the observed speech spectrum, while "formant" analysis attempts to identify and track the resonances explicitly.

Implementation of filter-bank and linear-prediction analysis methods is straightforward using modern digital signal processing, but suitable special-purpose or programmable components are only just becoming available at a price appropriate for the mass market. More sophisticated analysis methods will call for at least another order of magnitude in signal processing power, but this is expected to be available by the time we know how to use it.

32.5 Speech pattern processing – from patterns to words

Automatic speech recognition (ASR) is difficult. Among the generic problems are CONTINUITY of the sound pattern, VARIABILITY of the same word in different contexts or spoken by different persons, AMBIGUITY between patterns for different words, and the sheer COMPLEXITY of the whole, multi-level system of spoken language. One consequence of the complexity is that ambitious automatic speech recognition systems are often too complex for their developers to debug, let alone improve.

32.5.1 Isolated-word recognition using whole word models

The most popular and successful technique for ASR solves the above problems by re-defining ASR so that the problems are by-passed or minimised. Instead of trying to recognise anything, said by anyone, in a normal speaking style, the designers of the first commercially-available speech recognition machines insisted that: the set of words would be limited to a few dozen at any one time; the words must be uttered with distinct gaps of silence between them; the user of the machine must provide examples of all the words in an "enrolment" or "training" phase before the machine attempts recognition.

The first step is to turn each utterance into a pattern, using some form of short-term spectrum analysis, such as the filterbank analysis introduced above. It is also necessary to determine what part of the pattern of sound picked up by the microphone corresponds to the utterance to be analysed. This figure-ground separation, or endpoint detection, is often very difficult.

For each word, one or more "word models" are constructed from the training data. Word models can be simple single-example patterns ("templates") or sophisticated "stochastic models" (Baker and Baker [7]). One of the most serious types of variability among different utterances of the same word by the same speaker is variations in the timescale. The most successful isolated-word recognition machines incorporate a powerful method of comparing word-patterns which copes with unknown non-linear differences in timescale, and this method is even more important for connected word recognition. The mathematical basis of this "non-linear-timescale matching" is known as "dynamic programming".

Several ways of thinking about time-flexible pattern matching can lead to the same algorithm (Bridle [8]), but the most general derivation starts with a stochastic model (or "hidden Markov model") for each word. This "word machine" is supposed to go through a sequence of states, making transitions at regular intervals (transitions can cause states to be repeated). For each state transition, the machine produces an output which is an observable item such as a spectrum shape. The actual output for each transition is drawn from a distribution whose parameters are specific to the new state.

Given such a specification and a particular observed sequence of "outputs", we can compute the likelihood that any particular sequence of states migh occur and produce that observed sequence. (This is simply the product of the likelihoods of the supposed state transitions and the likelihoods of the observations being produced by the states associated with each of them.) The likelihood of the most likely sequence of states, given the observations, can be found very efficiently using a "dynamic programming" algorithm. A similar algorithm can compute the sum over all possible state sequences. Given an unknown speech pattern, we compute one of the above measures of fit of the data to the model for each word machine and decide on the most likely word.

In the case that the word-models are derived from single examples (or by simple averaging), then the assumption is usually made that the output distributions are isotropic multivariate normal, and the Euclidean distance squared to

Advanced signal processing of automatic speech recognition

the mean is therefore proportional to minus the log of the likelihood [8]. The maximum likelihood sequence of states corresponds to the timescale arrangement which minimises the squared error between the input and the word model (which is now referred to as a "template"). The dynamic-programming timescale-alignment and scoring method is sometimes known as "dynamic time warping".

The state transition probabilities model the flexibility of the timescale. It has been shown that the use of such information can dramatically increase recognition performance in the case of some very similar pairs of words (Russell *et al.* [9]).

32.5.2 Connected word recognition

It is possible to remove the restriction that the speaker must leave gaps between words. This allows faster data entry and needs less skill, but all the other limitations still apply. Perhaps the obvious approach is to divide the input pattern into words, then recognise each word as before. Unfortunately continuity beats us: consider "three eight" spoken fluently – the only significant discontinuity is in the middle of the word "eight".

Alternatively, we could try all sequences of words, concatenate the corresponding word models, compare them with the unknown speech pattern in a way that could cope with unknown differences in timescale, and choose the best-fitting sequence. Although it ignores many known phenomena of connected speech, this would give usefully good answers. However, it would take an impractical amount of computation.

The currently favoured approach, which is the basis of several commercially-available connected word recognisers [10, 11], does indeed find the sequence of word models which, as a single composite model, explains the whole of the input best. However, this is done without trying all possible sequences of words. Efficient connected-word recognition algorithms are again based on dynamic programming, and the solution to the template sequence problem can in fact be integrated into the solution of the timescale variability problem [12, 13]. In published work on recognition performance [14, 15] the best isolated word recognition performances have been achieved by connected word recognisers.

It is possible to take the integration of levels of the problem even further, and include a statistical model of language. This amounts to probabilities for transitions between complete-word models. Using this approach, some success has been reported for the "decoding" of sentences dictated from real documents (Bahl *et al.* [16]).

32.6 Future developments – from 1984 to 1994

This is an interesting time for speech technology research, because advances are being made in implementation technology faster than in our understanding of speech structure and discovery of effective algorithms for exploiting it. Consequently we are moving from a phase when we were constrained by the implementation technology to a new phase when we will be constrained by our ignorance of speech itself.

The easiest development to forecast is the implementation in low-cost hardware of the type of ASR capability which currently costs many thousands of pounds. This will be achieved by a combination of special-purpose integrated circuits and more effficient algorithms [1]. The result will be useful speaker-trained, small vocabulary, isolated (and limited connected-word) recognition for wristwatches, personal computers and the like.

The development of advanced stochastic modelling techniques currently in research laboratories will probably result both in small vocabulary, speaker-independent systems, (which will be very useful over the telephone) and also larger-vocabulary systems that can adapt to the characteristics of a new speaker without needing examples of all the words.

However, machines are likely to fall far short of human performance with large vocabularies and unknown speakers until it is possible to acquire and use far more knowledge of speech structure than seems possible with current recognition approaches. Several speech research laboratories are attempting to formalise speech structure knowledge at many levels, and hope to use a variety of "artificial" intelligence" techniques to apply this knowledge in automatic speech recognition.

Perhaps "real" automatic speech recognition will elude us until the pattern processing in our machines is equivalent to the perceptual processing performed by the brain. There has been significant progress in visual perception (Ballard et al. [17]), and some interesting indications of the nature of normal fluent human speech perception (Marslen-Wilson [18]).

References

1. WHITE, G. M., 1984, 'Speech recognition: an idea whose time is coming', *Byte*, **9**, pp. 213–222.
2. HOLMES, J. N., 1982, 'A survey of methods for digitally encoding speech signals', *The Radio and Electronic Engineer*, **52**, pp. 267–276.
3. REPP, B. H., 1981, 'On levels of description in speech research', *J. Acoustical Society of America*, **69**, pp. 1462–1464.
4. KOPEC, G. E., OPPENHEIM, A. V., and DAVIS, R., 1982, 'Knowledge-based signal processing', *Trends and Perspectives in Signal Processing*, **2**, pp. 1–6.
5. DUDLEY, H., 1940, 'The carrier nature of speech', *Bell System Technical Journal* **XIX**, pp. 495–515.
6. MARKEL, J. D., and GRAY, A. H., 1976, 'Linear Prediction of Speech', Springer-Verlag.
7. BAKER, J. M., and BAKER, J. K., 1983, 'Aspects of stochastic modelling for speech recognition', *Speech Technology*, **1**, No. 4.
8. BRIDLE, J. S., 1979, 'Pattern recognition techniques for speech recognition', *Spoken Language Generation and Understanding*, Nato Advanced Study Institute, Series C, D. Reidel Publishing Company, London, pp. 129–145.

9. RUSSELL, M. J., MOORE, R. K., and TOMLINSON, M. J., 1983, 'Some techniques for incorporating local timescale variability information into a dynamic time-warping algorithm for automatic speech recognition', Proc. IEEE Int. Conf. Acoustics, Speech and Signal Processing, Boston.
10. PECKHAM, J. B., GREEN, J. R. D., CANNING, J. V., and STEPHENS, P., 1982, 'A real-time hardware continuous speech recognition system', IEEE Int. Conf. Acoust., Speech Signal Processing, pp. 863–866.
11. WILSON, J., 1983, 'Connected speech system aims for industrial market', *Sensor Review*, **3**, No. 4.
12. VINTSYUK, T. K., 1971, 'Element-wise recognition of continous speech composed of words from a specified dictionary', *Kibernetika (Cybernetics)*, **7**, pp. 361–372.
13. BRIDLE, J. S., BROWN, M. D., and CHAMBERLAIN, R. M., 1983, 'Continuous connected word recognition using whole word templates', *The Radio and Electronic Engineer*, **53**, pp. 167–175.
14. DODDINGTON, G. R. and SCHALK, T. B., 1981, 'Speech recognition: turning theory to practice', *IEEE Spectrum*, **18**, pp. 26–32.
15. BAKER, J. M., 1981, 'How to achieve recognition: a tutorial/status report on automatic speech recognition', *Speech Technology*, **1**, pp. 30–43.
16. BAHL, L., COLE, A., JELINEK, F., MERCER, R., NADAS, A., NAKAMOO, D., and PICHENY, M., 1983, 'Recognition of isolated-word sentences from a 5000-word vocabulary office correspondence task', Proc. IEEE Int. Conf. Acoustics, Speech and Signal Processing, Boston.
17. BALLARD, D. H., HINTON, G. E., and SEJNOWSKI, T. J., 1983, 'Parallel visual computation', *Nature*, **306**, pp. 21–26.
18. MARSLEN-WILSON, W. D., 1979, 'Speech understanding as a psychological process', *Spoken Language and Understanding*, Nato Advanced Study Institute, Series C, D. Reidel Publishing Company, London, pp. 39–67.

Copyright © Controller HMSO, London 1985

Chapter 33
VLSI architectures for real-time image processing

Dr. P. M. Narendra
Honeywell Inc., Data Conversion Product Center, Colorado Springs, CO 80906, USA

33.1 Introduction

Image processing has left the labs for the field. Computer vision has gone through feasibility demonstration in a number of applications — visual inspection (1), (2) automatic assembly (3), and robotics (4), (5) and is ready for real-time cost effective implementations. This is especially true of military applications, in real-time scene analysis for the detection, recognition and tracking of targets in video from real-time (TV-like) sensors (6). Real-time implementation provides an exceptional challenge because of the extremely high data rates (in excess of 20 MHz), which result in computational throughputs of several thousand mega operations per second (MOPS). Even harder to realize are the requirements of low cost and size of the hardware integrated with the sensors.

It is clear that we need specially optimized architectures to achieve the throughputs required for real-time implementation, although most image processing algorithms have been developed on general purpose machines. Image processing algorithm researchers have paid scant attention to real-time implementation of the algorithms, focusing instead on non real-time simulation. Even where real-time implementation is mandated, the limitations of off-the-shelf components have severely restricted the algorithms implemented in prototype hardware (6).

VLSI offers the promise of achieving the requisite throughputs in reasonable size hardware. Now is the opportune figure moment to examine optimum computing structures for image processing algorithms to drive the development of VLSI devices. Successful real-time implementation with VLSI will require not only that the algorithm designer understand architectural considerations but also that the VLSI architect gain an intimate understanding of the image processing computational structures.

This paper atempts to classify both the inherent data flow and computational structures for a wide variety of image processing algorithms, and the candidate architectures suitable for VLSI implementation. The merits of each architecture are discussed against VLSI implementation criteria.

33.2 Image processing data flow and computational structures

A typical computer vision paradigm (with specific application to target detection and recognition in military environments) is shown in Figure 1. Typical sensor formats include TV and FLIR (forward looking infrared) which have either 525 ($\simeq 10$ MHz) and 875 line ($\simeq 20$–30 MHz) video format which is line scanned and has a 2:1 interlace at 30 frames per second. Several other sensor formats exist for specialized applications, including 1000-2000 pixel linear arrays in pushbroom and staring sensors (non-scanning) focal planes in sizes ranging from 64×64 to 256×256.

Common to all the sensors is the extremely high data rate and a serial format. VLSI architectures for real-time implementation have to take sensor and algorithm data flow into account in a topdown design methodology to prevent the classical I/O bottlenecks with plague bottom up implementations.

By far the highest throughput functions involve every pixel in the image at the input rate – 2D filters for image enhancement (contrast enhancement, high frequency emphasis filters, median filters, stochastic filters, etc.) and primitive image processing functions in scene analysis (edge operators, gradient, Hough transform, matched filters, convolutions and correlations). Following the high throughput functions in segmentation, the data rate falls subsequentially because of the need to process but only a limited number of discrete requirements, the segmentation function is a good candidate for VLSI implementation.

In spite of their apparent diversity, the multitude of image processing functions used in scene analysis fall into natural categories requiring common computational structures requiring different degrees of concurrency. In point, parallel type algorithms, every pixel in the image is processed independently of the neighboring elements, and is a function of the input pixel or corresponding pixels in two or more images. Examples of this function are the classical lookup table-type functions (histogram modification, grey scale transformations) addition and subtraction of two images, etc. Filtering in the time domain (over successive looks) at every pixel also falls in this category.

Another class of functions which require very high throughputs are sliding window functions, pervasive in image processing. Here, every output pixel is a function of the input pixels in a window surrounding the center pixel. The same function is computed for every pixel in turn. Sliding window functions can be classified as those requiring arithmetic operations (convolution, correlation, Sobel edge, etc.); those requiring compare or sort operations (medium filter, maximum suppression, etc.); and those involving Boolean operations on binary images (the endoskeleton, exoskeleton, perimeter, etc.).

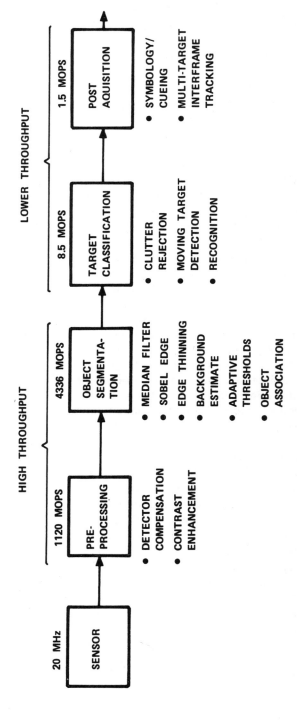

Fig. 33.1. *A computer vision Paradigm for a military application*

Most commonly occuring image processing functions can be cast into one of the above classes. Algorithms that do not directly fall under these classifications can often be transformed into one that will – for example, the Hough transform can be expressed as matched filters, which are window functions.

Therefore, we see that the generic VLSI architectures must optimize not only to the sensor data flow structures to minimize the input/output bottlenecks and buffering requirements, but also must have the computational structures to implement the above class of point transform and sliding window functions efficiently. Before examining the specific candidate architectures, however, let us establish the criteria for efficient VLSI implementation.

33.3 Criteria for VLSI implementation

There are several restrictions imposed by both the system and algorithm requirements and VLSI technology which must be used to judge candidate architectures. These are summarized below:

33.3.1 Computational throughput

This is the ability to meet the required computational throughputs with a given system clock rate. For example, VLSI clock rates (several gate delays) will range from 20 MHz for MOS technologies to several hundred MHz for bipolar technologies. Architectures which are inherently limited by the system clock rates are obviously undesirable. As a trivial example, a sequential microprocessor could not implement given the simplest of image processing functions at the 20 MHz input data rate. It is obvious that both pipelining and inherent parallelism of the image procesing operations will have to be exploited.

33.3.2 Programmability

All current high speed implementations of image processing functions tend to be hardwired (with off-the-shelf MSI and LSI logic). Programmability is an extremely desirable feature in VLSI, because it affords flexibility in implementing several different functions with the same chip set and the ability to change an algorithm in real-time adaptively to meet a scene dependent need.

33.3.3 Optimum computational structure

The architecture must be optimized to the image processing functions i.e. must be capable of implementing the typical computational structures like lookup tables and moving window functions efficiently. Unlike radar signal processing functions, however, full precision multiples and divides are rare. FFT's are typically not used because of the non-linearity of the functions the small size of the filters and the need for storing the entire image being processed. On the other hand, bandwidth compression of images does require transform domain processing which must be considered in deriving generic VLSI device architectures.

33.3.4 *Optimum data flow structure*

The architecture must be mated to the sensor data flow to facilitate direct interface to the real-time sensors. This is typically disregarded by architects who design special purpose hardware as peripherals to GP computers. If the architecture is not optimized to the data flow from the very beginning, inordinate amounts of buffering, format conversions, and SSI/MSI "glue" will be required, negating the benefits of integration.

33.3.5 *Modularity*

This is an absolute requirement for a general purpose VLSI chip to ensure a wide application base, i.e., can the same VLSI chip set meet the requirements of different sensor formats, data rates and algorithm needs? This means that the architecture must not only be modularly expandable to handle different sized focal planes (a 64×64 staring array vs. a 1024×1024 sensor) but also be capable of meeting a wide spectrum of functional throughputs and structures to meet different algorithmic requirements – a 3×3 Sobel edge operator, or a 15×15 median filter, etc. Ideally, modularity would allow the same VLSI chip set to achieve a variety of throughputs.

33.3.6 *Natural VLSI chip partitioning*

Can architecture be partitioned into VLSI chips with minimal chip to chip interconnects? Even with high density VLSI packages, I/O pins are at a premium and cause not only excessive power dissipation but also considerable off-chip delays, slowing the entire architecture down. In fact, some architectures may not even be partitionable into VLSI chips with reasonable pin counts unless the entire function is on one chip, as we will see subsequently.

33.4 Candidate VLSI architectures for real time image processing

This section will review candidate classes of architectures for real-time image processing in light of their applicability for VLSI implementation, according to the above criteria.

The high computational throughputs encountered in image processing (which exceed several thousand mega operations per second) demand concurrent processing. This concurrency can be exploited by either function (pipelining) or pixel-level parallelism. This leads to two broad classes of architectures: pipeline processors and SIMD parallel processing arrays. These are considered in turn.

33.4.1 *Serial pipelines*

In serial pipeline architectures, successive functions are performed on a serial data stream in cascaded stages with pipeline latches between stages. Since all the pipelined stages are operating concurrently, the total effective throughput is the sum of all the pipeline stages.

Serial pipelines for image processing accept the data in the line scan format and operate upon the input data as it is received. They are, therefore, optimized to the data flow of most imaging sensors, requiring no further scan conversion or buffering, and are economical in the storage required. Figure 2 shows an example of an implementation of a moving window function on a serial line scanned data format. The 1 H line delays shown are essentially first-in first-out (FIFO) buffers each storing one line of input, which can be implemented as either shift registers of RAMs. The line delays and the pixel delays (Δ) buffer the input data so that a moving window is proferred in parallel to the subsequent stage which must implement the function. For example, a 3 × 3 average would be computed as the sum of 9 elements. Remember that several operations have to be performed in one pixel duration. At the rate of 10–20 MHz pixel rates, this computational throughput can be extrememly high. Once the data has been staged, several options are feasible for implementation of the window function.

A sequential programmable architecture for the window function can be used but its main limitation would be the computational throughput. Even for a simple 3 × 3 averaging function, at 10–20 MGz, it would require instruction cycle speeds in excess of 100–200 MHz. Larger window functions and sort operations, for example, would not be implementable even at these clock rates.

Programmability of the window function can be achieved to a very limited extent with lookup table architectures. For example, the cytocomputer (7) uses a lookup table with 9 one-bit values as inputs from a 3 × 3 binary window. Its limitation is that a perfectly general function of a 3 × 3 window (8-bits wide) would require a lookup table RAM of 272 bits. It is possible to implement window funtions like the sum of products (convolutions) using lookup table based distributed arithmetic (8). Indeed, an LSI chip has been reported using this technique (9). Hoever, we are still limited to small window sizes (3 × 3) and by the access speeds of the lookup table RAM. For example, using the technique of (9), 9 memory references will be needed in each sample time. At 40 nanoseconds per pixel, this implies an extremely fast RAM (< 4 nsec).

Inner-product computers (10) are programmable but inherently lack the speed required for implementation of image processing functions. Further, typical image processing functions are not multiply-oriented but are instead add, subtract and compare-and-exchange dominant (edge operators, median filter, etc.).

It appears that the only serial pipeline solution to performing a general purpose window function on a sufficiently large size window is to have a hardwired pipelined architecture within the chip. An example would be a tree of adders to perform the sum of a 3 × 3 elements in three stages. Pipelining would allow each adder stage to perform at the pixel rate and achieve the necessary throughput within the chip. Systolic arrays (11) offer a regular implementation of the inner product-type functions with a small set of compu-

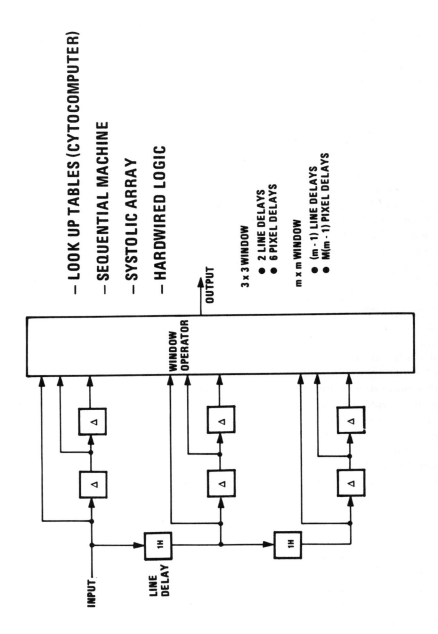

Fig. 33.2. *Hardware pipeline for image processing. Examples: Cyto-computer, Sobel edge chip, etc.*

tational cells which may reduce the chip design and lay-out cost. Similar VLSI architectures are feasible (12) for implementation of sort functions required in median filters, for example.

The main drawback of hardwired functional pipelines, of course, is that they are not programmable. This implies that different VLSI chip sets or at least configurations will be needed to perform different algorithms. Further, modularity suffers because of the need to customize the architecture for a given size window.

This architecture has the benefit that most of the storage required for buffering the moving window can be implemented off chip, exploiting commercial RAM developments. On the other hand, this partitioning raises the spectre of too many pins i.e. 120 for a 15 × 15 window function (if the pixel delays were on-chip). A further issue is partitioning into more than one VLSI chip if the entire function cannot be accommodated on one chip. This, of course, would depend upon the specific function being performed. A natural partition is not readily apparent.

The utility for VLSI must be examined in the light of programmability and modularity. Sometimes (7), it is possible to decompose image processing algorithms into several cascaded stages of elementary 3 × 3 window functions. However, this may be too confining a constraint for most image processing algorithms.

33.4.2 Two-dimensional SIMD parallel processing arrays

Two-dimensional single instruction multiple data (SIMD) parallel processing architectures have long been advocated for image processing (13), (14). Most recent of these are the distributed array processor (DAP 15)), and the massively parallel processor (MPP 10)), both of which have been suggested for image processing. Although currently being built of SSI and MSI components, these architectures are candidates for real-time VLSI implementations.

In (15) and (16), a two-dimensional array of simple parallel processing elements (in both cases employing simple bit-serial arithmetic) are connected with four nearest neighbor interconnections (see Figure 3). Each processing element is associated with its own memory and paths are provided for getting the data into the array for processing and outputting the results after processing. The array and memory address control is broadcast to the processing element array and shared by the array. All PEs operate identically on their data concepts except when inhibited by an activity bit.

This architecture is ideal for the general moving window algorithms. For example, if each pixel were mapped to one processing element, the average of a 3 × 3 window can be expressed as performing accumulation in each processing element of the eight neighbors sequentially. At the end of eight operations each processing element would contain the result. Of course, depending upon the total throughput required, more than one pixel would be mapped to each processing element, and processed sequentially. Larger windows can be implemented by successive memory-to-memory transfers between connected PEs.

This architecture meets the criteria for the total computational throughput by taking advantage of the inherently parallel nature of the algorithms. (For example, a 1000 × 1000 array of procesing elements would have a throughput in excess of 10^{12} one-bit adds per second even at a system clock of 1 MHz if bit serial arithmetic is adopted). It also has the advantage of programmability

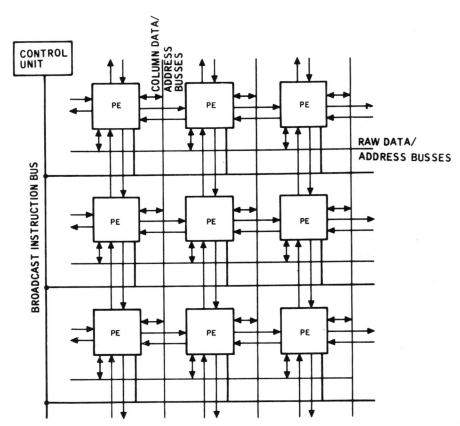

Fig. 33.3. *A two-dimensional SIMD array for image processing is typified by Massively Parallel Processor and DAP*

because the controller can be programmed to implement a sequence of different algorithms upon the data without the data leaving the array. It meets the VLSI criteria because the processing elements are identical and make the design and layout of VLSI chips, testablity and fault more feasible. This architecture is ideally suited for temporal processing i.e. processing in time at each pixel position (for example, time average over several frames).

The principal drawbacks of the 2D SIMD parallel arrray architecture are that while it is optimized to the computational structure of the window algorithms, it is not tailored to the data flow from the sensor. It is easy to see that when operating upon the sequential input (line scan), the array will need to buffer the entire image before processing can begin. If the processing has to occur concurrently with the input and if sequential output is desired, three full frames of data have to be buffered for input, processing, and output. This may be a small overhead for a function (or functions) requiring very high throughputs i.e. where processing dominates input/output. For small throughput functions, however, this may be a major overhead. For example, a 3 × 3 Sobel edge operator operating upon a sequential input, would only require two lines of storage for a serial pipeline implementation, while the 2D array would need to buffer the entire image before proceeding to compute the output. Getting sequential data into and off the parallel array needs careful addressing for real-time implementation because the current architectures (15), (16) do not address this issue. MPP allows the shifting in of a bit plane of image concurrently with the process, but it would still require a format conversion buffer to interface with a sensor.

The partitioning of this architecture into VLSI chips requires the addressing of two issues – the number of pin-outs needed within each VLSI chip; and the on-chip vs. off-chip memory tradeoffs. The off-chip memory for reasonable size PE arrays will entail a very high I/O bandwidth and a large number of I/O pins. For example, a 16 × 16 bit serial array would have 512 I/O pins and corresponding high driver power for off-chip memory. This chip would require 64 input and 64 output pins for neighbor communication as well. Moreover, increasing gate density on-chip increases the number of interconnects correspondingly.

33.4.3 *MIMD architectures*

Multiple Instruction Multiple Data (MIMD) architectures have been proposed for real-time image processing because they offer more flexibility for adaptive algorithms. These MIMD architectures are loosely coupled bus oriented or network connected distributed architectures. Bussed architectures do not have the bandwidths required for interprocessor I/O in the algorithms described above. In switched network architectures, and the control and bus management overhead may be too large a price to pay for the flexibility. Also, because control is not shared among the large number of processing elements, the MIMD architecture may not be efficient for the regular high throughput image processing functions at the front end of computer vision systems. On the other hand, multi-processor distributed architectures may indeed be suitable for the subsequent higher level functions after the data rates have been reduced through array feature extraction. Here the programmability afforded by loosely coupled distributed processors is not only desirable but necessary.

33.5 Summary

In this paper, we have attempted to bring together the algorithm and data flow requirements of real-time image processing functions with the constraints for VLSI implementations to perform a comparative evaluation of architectures. This evaluation is of necessity subjective and serves mainly to highlight the key differences in the architectures and the computational structures of the algorithms.

The basic message is that the image processing algorithm and system designer has to understand the VLSI architecture drivers; the VLSI architect has to address the data flow and computation structures for image processing functions. Honeywell is developing programmable VLSI devices which are simultaneously optimized to the sensor data flow and the algorithm structures while processing flexibility to meet a number of system requirements.

33.6 Acknowledgement

The author would like to acknowledge Sharon McDaniel and Julie Linley's excellent assistance in preparing this paper.

33.7 References

1. JARVIS, J. F., 1977, 'Automatic Visual Inspection of Western Electric Series 700 Connectors', *Proc. PRIP77, IEEE Comp. Soc. Conf. Patt. Rec. and Im. Proc.*, pp. 153–159.
2. HARLOW, C. A., *et al.*, 1975, 'Automated Inspection of Electronic Assemblies', *Computer*, **8**, pp. 36–45.
3. BAIRD, M. L., Feb. 1978, 'SIGHT-1: A Computer Vision System for Automated Chip Manufacture', *IEEE Trans. Systems, Man and Cybernetics*, **SMC-8**, pp. 133–139.
4. WARD, M. R., *et al*, 1979, 'Consight: A Practical Vision and Sensor Based Robot Guidance System', in *Computer Vision and Sensor Based Robots*, pp. 81–100.
5. HOLLAND, S. W., Aug. 1977, 'An Approach to Programmable Computer Vision for Robots', *Report No. GMR2519*, General Motors Research Laboratories.
6. SOLAND, D. E. and NARENDRA, P. M., April 1979, 'Prototype Automatic Target Screener', *Proceedings SPIE, Smart Sensors*, **178**, pp. 175–184.
7. LOUGHEED, R. M., *et al.*, 1980, 'The Crytocomputer A Practical Pipelined Image Processor', *Proceedings of the 7th Annual International Symposium on Computer Architecture*.
8. BURRUS, S. C., December 1977, 'Digital Filter Structures Described by Distributed Arithmetic', *IEEE Transactions on Circuits and Systems*, pp. 674–680.
9. DEMAN, H. J., *et al.*, October 1978, 'High Speed NMOS Circuits for ROM-Accumulator and Multiplier Type Digital Filters', *IEEE J. Solid-State Circuits*, **13**, pp. 565–572.
10. SWARTZLANDER, E. G., *et al.*, January 1978, 'Inner Product Computers', *IEEE Trans. Computers*, **C-27**, pp. 21–31.
11. KUNG, H. T., July 1980, 'Special Purpose Devices for Signal and Image Processing', *SPIE Proceedings*.
12. NARENDRA, P. M., May 1978, 'A Separable Median Filter for Image Noise Smoothing',

Proc. 1978 Conf. Patt. Rec. and Image Proc., pp. 137–141. Also, to appear in IEEE Trans. Pattern Analysis and Machine Intelligence, January 1981.
13. McCORMICK, B. H., Dec. 1963, 'The Illinois Pattern Recognition Computer, ILL1AC III', *IEEE Trans. Computers*, pp. 791–813.
14. KRUSE, B., April 1975, 'A Parallel Picture Processing Machine', *IEEE Trans. Computers*, **C-14**, pp. 424–433.
15. GOSTICK, R. W., May 1979, 'Software and Algorithms for the Distributed-Array Processors', *ICL Technical Journal*, pp. 116–135.
16. BATCHER, K. E., October, 1979, 'Massively Parallel Processor (MPP) System', *AIAA Computers in Aerospace Conference II*, Los Angeles, CA.

Index

actively matched amplifiers, 60
acoustic devices, 253
acoustic-optic devices, 93
acoustic-optic correlators
 – one dimensional, 95
 – two dimensional, 97
ADA, 127, 242
adaptive moving–target indicator, 224
adaptive CFAR, 226
aerials, 1
airbridge, 57
alpha-numeric labels, 175
Alvey programme, 118
{ ambiguity, 276
 ambiguity diagram, 92
 ambiguity–function processor, 98
analog-to-digital converter, 103, 226, 253
antenna, 1
array, 1, 25
array antenna processing, 223
artificial intelligence, 271
ASARS system, 269
automatic speech recognition, 275

baffle, 26
bats, 91
beamforming, 85
bipolar transistors, 10
bilinear transformation, 146
bit-error rate, 37
Bragg cell, 94
bulk-effect devices, 16
buffered fet logic, 106
built-in test, 133

c.a.d. interface standards, 130
carrier concentration, 53
cathode-ray tube
 – cursive displays, 154
 – raster displays, 155
c.c.d. shift registers, 94

cell library, 112
ceramic chip carriers, 230
channel thickness, 53
channel pinch-off voltage, 106
charge-coupled device, 84, 93, 143, 252
charge-transfer inefficiency, 88
chirp-z transform, 83, 88
clutter, 4
coherent systems, 39
coherent solid-state light sources, 93
colour, 173
computer-aided design, 114, 116, 128, 141, 192, 239
complex multiplier, 227
complexity, 277
connected-word recognition, 279
constant current, 9
constant power, 9
constant voltage, 9
constant false-alarm rate, 224, 262
contact printing, 108
continuity, 277
control interface, 234
convolution, 69, 219, 233, 249, 283
co-ordinate transform, 233
correlation, 69, 93, 233, 251, 283
couplers
 – directional, 207
 – block, 207
cross-ambiguity function, 98
cross correlation, 21
cross-field amplifier, 16
cryogenic sensors, 248
custom l.s.i. design, 114

dark current, 97
declarative methods, 266
design automation, 141
differential phase-shift keying, 39
digital CFAR processor, 229
digital communications, 126

Index

digital delay line, 93
digital filters, 143
digital pulse compression, 226
digital signal processing, 126, 138, 141, 253
dipod, 124, 274
directivity, 3
discrete convolution, 83
discrete Fourier transform, 83
distributed-array processor, 123, 229, 257, 289
diversity reception, 40, 42
diversity techniques, 35
domes, 24
dolphins, 91
Doppler processing, 91, 226
dynamic pattern adaptivity, 225
dynamic programming, 276, 278
dynamic time warping, 279

electroluminescent display
 – a.c. approach, 156
 – d.c.-powder approach, 157
electromagnetic compatibility, 15, 18
electromagnetic simulators, 183
electromagnetic trials ranges, 183
electronic support measures, 185
emulation, 181
emulation machines, 125
endoskeleton, 283
energy buffer, 11
environment, 182
equal-gain combining, 45
ergonomics, 170
exoskeleton, 283

fast Fourier transform, 219, 240, 232, 254, 260
fault tolerance, 141
feedback, 183
feedforward regulation, 13
FFT address sequencer, 228
FFT butterfly, 228
FFT kernel, 219
finite wordlength, 149
Fith, 124
flash converter, 103
flicker-free images, 166
flow noise, 26
format analysis, 276
FORTRAN, 258
frequency distortion, 19
functional simulators, 113
functional throughput rate, 120
functional units, 233

GaAs mesfet, 103
GaAs fet comparator, 104
gallium-arsenide integrated circuits, 50, 199
gain matching, 77
gas discharge display
 – a.c. plamsa panel, 157

 – d.c. gas discharge panel, 158
gate array, 112, 128
gate length, 53
GRID processor, 218
group delay, 72, 77

high-throughput processing systems, 193
Hough transform, 283
human efficiency, 177
hybrid signal processing, 256
hydrophone, 25
hydrostatic pressure, 26

image processing, 217, 282
image understanding, 265
impedance matching, 11
integrated design automation, 138
intelligent knowledge based systems, 119
interdigital capacitors, 54
interferometer, 27, 28
intermodulation distortion, 55
intermodulation products, 68, 84
isolated-word recognition, 278
isomorphic transformation, 265

Josepheson technology, 253
just-noticeable differences, 166

klystron, 16

Lange quadrature couplers, 54
laser diode, 204
lighting, 166, 172
limiting, 76, 77
linear-frequency modulation, 92
linear-period modulation, 92
linear-phase filter, 148, 151
linear-prediction analysis, 277
liquid crystal display
 – guest host, 159
 – twisted nematic, 159
logarithmic amplifiers, 72
 – parallel summation, 76
 – series summation, 76
 – successive approximation, 72
 – true logarithmic, 72
 – video amplifiers, 72
logic simulators, 113
log-modulus extractor, 228
l.s.i., 110, 133, 233

macrocell, 127, 128
macro systems, 197
magnetron, 16
maintainability, 18
master control unit, 259
matching inductors, 56
matrix multiplication, 206

Index

MASCOT, 196
massively parallel processor, 218
maximal ratio combining, 43
maximum entropy, 224
maximum likelihood, 224
mechanical scanning, 161
mesfet, 51
modulation
— low frequency, 19
— within-pulse repetive, 20
monolithic microwave integrated circuits, 50
mosfets, 10
mos shift registers, 94
motion, 182
multilayer ceramic, 32
multiple-instruction multiple data processors (MIMD), 195, 291
mutual coupling, 9

noise figure, 53, 78
non-coherent frequency-shift keying, 38
non-recursive filter, 144, 147
number-theoretic transforms, 199

OLYMPUS satellite systems, 209, 211
optical devices, 93, 253
optical fibres, 26, 203
optical photolithography, 108
overlay capacitors, 54

parallel microprogrammed processor, 229
pattern recognition, 126
perceptual processing, 279
perceptual structuring, 174
perimeter, 283
PERQ, 261
phase-shift keying, 214
photodiode arrays, 95
piezo-electric, 2, 32
pin grid arrays, 230
pipeline processors, 286
polycell design, 112
POP, 269
power converter, 15
processing element, 218, 229, 259, 289
processing gain, 249
programmable logic array, 112, 127
programmable processing
— matched filters, 25, 254
— pulse compression, 223
— spatial, 80, 81
programmable processing
— signal processing, 80, 82
— transversal filter, 86
PROLOG, 267
propagation delay, 77
proton exchange, 63
pulse compression, 2, 4, 83, 250
pulse expansion, 4, 250

pulse-width modulation, 10

quadrature phase-shift keying, 214

Raman amplifiers, 208
Rayleigh distribution, 36
recursive filter, 145
recursive processors, 199
reliability, 18
reflectometric sensor, 28
reverberation, 4
rounding noise, 149

sample and hold, 108
satellite systems, 209
Schottky barrier voltage, 106
scheduler, 243
security, 116
selection diversity, 43
semi-custom i.c. design, 111
shape, 175
sidescan sonar, 161
signal synthesiser, 15
signal processing graph notation, 196
silicon photodiode arrays, 93
simulators, 180, 239
simulator elements, 180
single-instruction multiple-data processors, 195, 286, 289
Sobel edge, 283
spatial spreading, 4, 5
speech waveforms, 274
speech pattern processing, 277
speech perception, 276
SPICE, 76
spread-spectrum processing, 96, 215
standard cell, 112
start-up delay, 15, 18
state equations, 143
state variables, 143
statistical clustering, 266
sticks diagram, 114
stimulation, 179
stochastic modelling, 280
submicrometer technology, 137, 198
superconductive circuits, 248
surface-acoustic wave filters, 143
sustainer, 159
switched-capacitor filters, 143
symbol processing, 276
syntatic analysis, 267
systolic processors, 195, 228

tangential sensitivity, 78
technology insertion, 137
test sequences, 113
time-division multiple access, 214
time spreading, 4
transducer, 1

travelling-wave amplifiers, 60
travelling-wave tube, 16
transmitter, 1
transversal filter, 85, 144
trials simulators, 187

u.l.s.i., 192
uncommitted-logic array, 112
uniform distribution, 36
UNIX, 262

vacuum florescent displays, 155
vapour hydride, 108
variability, 277
VHPIC, 88, 118, 192, 222
 – process performance, 121
 – system performance, 121
VHPIC application demonstration
 programme, 119
VHPIC d.a.p., 124
VHPIC design, 127
VHSIC, 88, 118, 132, 192

– brassboards, 132
– program status, 132
– program structure, 134
– phases, 134
– hardware description
 – language (VHDL), 141
VISIONS system, 268
visual coding, 175
virtual machines, 125
v.l.s.i., 110, 119, 133, 218, 232, 263, 282
vocoders, 277
voltage standing-wave ratio, 51

wave digital filter, 146
wide-band noise, 21
within-pulse scanning, 161
workspace, 171

yield enhancement, 137

z-transform, 143